HARDY SPACES ON HOMOGENEOUS GROUPS

by

G. B. Folland and E. M. Stein

Mathematical Notes 28

Princeton University Press

and

University of Tokyo Press

Princeton, New Jersey

1982

Published in Japan exclusively by University of Tokyo Press;
in other parts of the world by Princeton University Press.

Library of Congress Cataloging-in-Publication Data

Folland, G. B.
Hardy spaces on homogeneous groups.
(Princeton mathematical notes ; 28)
Bibliography: p.
Includes indexes.
1. Hardy spaces. 2. Functions of real variables. 3. Lie groups. I. Stein, Elias M., 1931-
II. Title. III. Series.
QA331.5.F64 515.7'3 82-47594
ISBN 0-691-08310-X (pbk.) AACR2

Printed in the United States of America

TABLE OF CONTENTS

INTRODUCTION

The object of this monograph is to give an exposition of the real-variable theory of Hardy spaces (H^p spaces). This theory has attracted considerable attention in recent years because it led to a better understanding in $\mathbb{R}^n$ of such related topics as singular integrals, multiplier operators, maximal functions, and real-variable methods generally. Because of its fruitful development it seems to us that a systematic exposition of some of the main parts of this theory is now desirable. There are, however, good reasons why in addition the theory should be recast in the more general setting where the underlying $\mathbb{R}^n$ is replaced by a homogeneous group. The justification for this wider scope, both in terms of the structure of the theory and its applications, will be described in more detail below.

Background: Development of H^p theory

The theory of H^p spaces is a multifaceted one with a rich history. Here we can do no more than sketch its highlights*.

Originally H^p $(0 < p < \infty)$ was defined to be the space of holomorphic functions F on the unit disc or upper half-plane such that

---

(*) See also the overview by C. Fefferman [2].

$$\sup_{0<r<1} \int_0^{2\pi} |F(re^{i\theta})|^p d\theta < \infty \qquad \text{(for the disc)}$$

$$\sup_{0<y<\infty} \int_{-\infty}^{\infty} |F(x+iy)|^p dx < \infty \qquad \text{(for the half-plane).}$$

The theory of these spaces (which began more than sixty years ago) developed as an important bridge between complex function theory and Fourier analysis. On the one hand methods of complex function theory such as Blaschke products and conformal mappings played a decisive role; on the other hand use of these tools yielded deep results in Fourier anlysis. Accounts of this part of H^p theory can be found in Zygmund's treatise [1], and the books of Duren [1] and Koosis [1].

It was natural that extensions of this theory would be sought with $\mathbb{R}^n$ replacing the real-line or the unit circle. One was therefore led to consider systems of harmonic functions on $\mathbb{R}^n \times (0,\infty)$, satisfying certain natural generalizations of the Cauchy-Riemann equations there, with the H^p condition being

$$\sup_{0<y<\infty} \int_{\mathbb{R}^n} |F(x_1,\ldots,x_n,y)|^p dx_1 \cdots dx_n < \infty .$$

In this theory of harmonic H^p spaces it was also natural to shift the focus of attention from the harmonic functions themselves to their boundary values, which are distributions on $\mathbb{R}^n$ from which the harmonic functions can be recovered by the Poisson integral formula. The resulting spaces $H^p(\mathbb{R}^n)$ are equivalent to $L^p(\mathbb{R}^n)$, when $p > 1$; but for $p \leq 1$ interesting results could be proved for H^p which did not hold for L^p. While complex function theory was not available, harmonic majorization proved to be at least a partial substitute. The approach we have just described originated about twenty years ago in the work of Stein and Weiss [1] (see also [2]).

Several years later the picture became clearer when it was found that the standard singular integrals and multiplier operators were in fact bounded on the H^p spaces, (see Stein [2]). Out of this emerged the point of view that $H^p(\mathbb{R}^n)$ was, as far as large parts of harmonic analysis were concerned, the appropriate substitute for $L^p(\mathbb{R}^n)$, when $p \leq 1$.

The years 1970-72 saw a series of rapid-fire developments which led to a profound reshaping of the theory. First, methods of Brownian motion were used (by Burkholder, Gundy, and Silverstein [1]) to show that $F = u+iv$ belongs to the classical H^p if the non-tangential maximal function of its real part u belongs to L^p. Thus while H^p theory was still tied to harmonic functions, it was freed (at least in one variable) of its dependence on the notion of conjugacy of harmonic functions. But two questions were implicitly raised: how could these results be extended to n-dimensions, and what was the rôle, fundamental or merely incidental, of the Poisson integral in these matters.

The answers to these questions came quickly. These results were indeed extendable to n-dimensions, but more importantly, the definition of H^p was largely independent of the kind of approximate identity one used. Thus H^p could be defined in terms of any approximate identity (fashioned out of a fixed $\phi \in S$, with $\int \phi \neq 0$), or one could also take into account "all" possible approximate identities in terms of a very useful tool — the "grand maximal function". Also singular integrals and maximal functions were now placed on the same footing: very roughly speaking, H^p was characterized as being stable with respect to either class of operators. Thus H^p theory which was conceived and grew under the profound influence of complex analysis had now fully emerged as a proud standard-bearer for real-variable theory.

There were at that time several other striking results which confirmed this view. Firstly, the well-known duality of C. Fefferman between H^1 and the space of functions of bounded mean oscillation. Also the introduction of the function $f^{\#}$ which was used in the study of several types of operators whose L^p properties had hitherto proved resistant. As an example we mention that one has recently been able to prove a whole list of new sharp estimates for several versions of the wave equation*.

(*) The results for H^p described in the last two paragraphs are in C. Fefferman and Stein [1]. Estimates related to the wave equation may be found in Marshall, Strauss, and Wainger [1], and Peral [1].

A further real-variable insight into the theory came several years later (in the works of Coifman [1] and Latter [1]) when it was proved — using the grand maximal function — that elements of $H^p(\mathbb{R}^n)$, $p \leq 1$ had "atomic decompositions"; i.e. they could be built out of simple constituents. These "atoms" are, apart from normalizing factors, compactly supported functions with a certain number of vanishing moments. This decomposition turns out to be a technical tool <u>par</u> <u>excellence</u> since several important assertions for $H^p(\mathbb{R}^n)$ could be reduced to easy verifications for atoms. These facts led Coifman and Weiss [2] to define Hardy spaces in the context of "spaces of homogeneous type" in terms of atomic decompositions. For certain parts of H^p theory the results previously obtained were given a larger scope. At about the same time another extension was initiated, which while not as general as the one growing out of abstract atomic decompositions, did however have many parallels with the $\mathbb{R}^n$ theory and will be of particular interest to us. The context was $\mathbb{R}^n$, but the usual dilations were replaced by non-isotropic ones*.

<u>The rôle of homogenous groups</u>

A homogenous group (of which $\mathbb{R}^n$ is the simplest example) is a Lie group equipped with an appropriate family of dilations. That these groups form a natural habitat for extensions of many of the objects studied in

(*) See Calderón and Torchinsky [1], [2].

Euclidean harmonic analysis can be understood from several points of view. First, it is a trite observation that many of the basic operators of harmonic analysis in $\mathbb{R}^n$, and in particular in $H^p(\mathbb{R}^n)$ theory, are of the convolution type; and in addition these have some natural invariance with respect to dilations. If we seek to generalize this situation we are naturally led to groups which have appropriate one-parameter groups of automorphisms acting like dilations. These are the homogeneous groups.

More convincing than these *a priori* considerations are the occurences of homogeneous groups in many interesting applications, to wit: in the Iwasawa decomposition of semi-simple Lie groups and hence as "boundaries" of the associated symmetric spaces; also the rôle that they play in complex analysis in several variables and in some non-elliptic partial differential equations*. Let us elaborate these points. In the Iwasawa decomposition for the semi-simple group $G = KAN$, the nilpotent group N is homogeneous, where the dilations come from an appropriate one-parameter sub-group of A. The symmetric space G/K has as its "boundary" a homogeneous group $\bar{N}$ isomorphic with N; and harmonic functions on the symmetric space are in principle representable as Poisson integrals which are convolutions on the group $\bar{N}$ **. The boundary behavior of harmonic functions (e.g. Fatou's theorem)

(*) This was the point of view in Stein [4].

(**) See the survey of Koranyi [2].

and the corresponding maximal functions have, in analogy with $\mathbb{R}^n$, close connection with H^p theory on $\overline{N}$. Similarly this H^p theory is related with the homogeneous singular integral operators on $\overline{N}$; incidentally substantial examples of such operators arise as intertwining operators for the representations of G. (See Knapp and Stein [1].)

In the particular case when G is the group $SU(n+1,1)$ the symmetric space is the unit ball in $\mathbb{C}^{n+1}$, and the homogeneous group $\overline{N}$ which gives the boundary is the Heisenberg group H_n. This indicates a close connection between complex analysis and homogeneous groups. Indeed a generalization of H^p theory to the Heisenberg group, related as it is to complex analysis in the ball, has been carried out by Geller [2] and Latter and Uchiyama [1]. Moreover analysis on the Heisenberg group has intimate ties with the $\overline{\partial}$ complex on boundaries of strictly pseudo-convex domains (see Folland and Stein [1]). An impetus for further study of homogeneous groups was the link forged in Rothschild and Stein [1] between these groups and hypoelliptic differential operators(*). An interesting class of homogeneous groups which have many features of the Heisenberg group are the "stratified" groups studied by Folland [1], [3], and Saka [1], where various function spaces have been analyzed in anology with their Euclidean counterparts.

(*) See also the expository accounts of Folland [2] and Goodman [1].

Description of this monograph

Since it is our purpose to develop the theory of H^p spaces in the setting of homogeneous groups, it might be well to explain to the reader what are the main difficulties of this task. First, some parts of H^p theory can be generalized from $\mathbb{R}^n$ to homogeneous groups in a purely routine fashion; in others the basic ideas still work but are technically more difficult to execute; and yet other parts require genuinely new ideas, because two main pillars of H^p theory on $\mathbb{R}^n$ — the Fourier transform and Green's theorem — are largely unavailable in the general context. Examples of results where the techniques used are in part quite different from the analogous situation in $\mathbb{R}^n$ are: (i), the passage from special approximate identities to the grand maximal function in Chapter 4; (ii), the characterization of H^p by square functions, and in particular the converse direction; iii), a variant of the Marcinkiewicz multiplier theorem in the setting of functions of the sub-Laplacian for stratified groups in Chapter 6[*]; and (iv), the H^p maximal theorem which corresponds to restricted convergence of Poisson integrals to their boundary values on a general symmetric space, which is in Chapter 8.

Our general objectives have unfortunately forced us to forgo some interesting special topics, such as: the recent applications of H^p theory

---

[(*)] This result is joint work with A. Hulanicki. See the reference below.

to several problems related to function algebras*, and the discussion of topics related to conjugacy and special operators characterizing H^p.

The monograph is organized as follows. In chapter 1 we collect all the necessary background material. The basis of H^p theory is developed in Chapters 2 and 3, culminating in Theorem 3.30, which gives the equivalence of various characterizations of H^p in terms of grand maximal functions and atomic decompositions. The remaining five chapters all depend on this material but are for the most part independent of one another. The main exceptions are that the results of Chapter 4, dealing with relations among maximal functions, play a significant rôle in Chapter 8 in the study of boundary value problems; similarly the results on convolution operators in Chapter 6 are used crucially in Chapter 7 in the treatment of the square functions which characterize H^p.

At the end of each chapter we have included some notes and references in which we list our sources as well as related papers. We have tried to provide fairly complete references to work done since 1970, and we apologize ahead of time if any serious omissions have crept in.

This work was begun when the first author enjoyed the hospitality of the Institute for Advanced Study. The research of both authors was partially supported by the National Science Foundation. It is with pleasure that we also express our appreciation to Miss Perine Di Verita for her excellent typing of the entire manuscript.

(*) In this connection, cf. Koosis [1].

Remarks on Notation

As a possible help to the reader we present here a brief dictionary of some of the notation when $G = \mathbb{R}^n$, and the dilations are given by the usual scalar multiplications.

(i) group multiplication is then vector addition.

(ii) Δ is the set of non-negative integers.

(iii) for the geometrical constants, $Q = n$, and $\gamma = \beta = d_j = \bar{d} = 1$.

(iv) for derivatives, $X_j = Y_j = \frac{\partial}{\partial x_j}$, and if I is a multi-index $X^I = Y^I = (\frac{\partial}{\partial x})^I$, and $d(I) = |I|$.

See also the more complete Index of Notation on p. 283.

CHAPTER 1

Background on Homogeneous Groups

In this chapter we develop the background material that will be used throughout this monograph. Here is a brief description of what follows. Section A is devoted to the basic structural facts about homogeneous groups. Such groups are Lie groups, each endowed with an appropriate family of dilations. These groups are automatically nilpotent, and enjoy several other special characteristics. The other sections of this chapter are then in large part concerned with explicating the analogues for these groups of properties that are standard for $\mathbb{R}^n$ (with the usual dilations); as such these sections may be omitted in a first reading.

At the outset, let us establish a few basic notational conventions.

$\mathbb{R}$, $\mathbb{C}$, $\mathbb{Z}$, and $\mathbb{N}$ will denote the sets of real numbers, complex numbers, integers, and nonnegative integers. If $x \in \mathbb{R}$, $[x]$ will denote the greatest integer in x. The word "function" will always mean "complex-valued function" unless otherwise specified. If X is a topological space and $E \subset X$, the closure of E will be denoted by $\overline{E}$ and the complement of E by either $X \setminus E$ or E^c. The characteristic function of the set E will be denoted by χ_E. The letter C will be frequently used to denote miscellaneous constants whose precise value is irrelevant. Finally, the symbol # will be used to signal the end of a proof.

A. Homogeneous Groups

Let $\mathfrak{g}$ be a Lie algebra (always assumed real and finite-dimensional), and let G be the corresponding connected and simply connected Lie group. If U,V are subsets of $\mathfrak{g}$ (resp. G), we denote by [U,V] the subspace of $\mathfrak{g}$ (resp. subgroup of G) generated by the elements of the form $[\alpha,\beta]$ (resp. $\alpha\beta\alpha^{-1}\beta^{-1}$) with $\alpha \in U$, $\beta \in V$. The lower central series of $\mathfrak{g}$ and G are defined inductively by

$$\mathfrak{g}_{(1)} = \mathfrak{g}, \quad \mathfrak{g}_{(j)} = [\mathfrak{g},\mathfrak{g}_{(j-1)}]; \qquad G_{(1)} = G, \quad G_{(j)} = [G,G_{(j-1)}].$$

The $\mathfrak{g}_{(j)}$'s (resp. $G_{(j)}$'s) are a descending chain of ideals of $\mathfrak{g}$ (resp. normal subgroups of G), and it is known (cf. Bourbaki [1], III. 9.2) that each $G_{(j)}$ is a connected Lie subgroup of G whose Lie algebra is $\mathfrak{g}_{(j)}$. $\mathfrak{g}$ (or G) is called nilpotent if $\mathfrak{g}_{(m+1)} = \{0\}$ (equivalently, $G_{(m+1)} = \{id\}$) for some $m \in \mathbb{N}$; more precisely, if $\mathfrak{g}_{(m+1)} = \{0\}$ but $\mathfrak{g}_{(m)} \neq \{0\}$ we say that $\mathfrak{g}$ (or G) is nilpotent of step m.

We recall the Campbell-Hausdorff formula: if $\exp : \mathfrak{g} \to G$ is the exponential map, for $X,Y \in \mathfrak{g}$ sufficiently close to 0 we have

$$(\exp X)(\exp Y) = \exp H(X,Y) \tag{1.1}$$

where H(X,Y), the Campbell-Hausdorff series, is an infinite linear combination of X and Y and their iterated commutators. (More precisely, (1.1) is valid whenever the series H(X,Y) converges.) The exact formula for H may be found in Bourbaki [1], II. 6.4; for our purposes it will suffice to note that H is universal, i.e., independent of $\mathfrak{g}$, and that

$$H(X,Y) = X + Y + \tfrac{1}{2}[X,Y] + \cdots$$

where the dots indicate terms of order ≥ 3.

Let V denote the underlying vector space of $\mathfrak{g}$. If $\mathfrak{g}$ is nilpotent, the Campbell-Hausdorff series terminates after finitely many terms and defines a polynomial map from $V \times V$ to V. Moreover, this map is a group law which makes V into a Lie group whose Lie algebra is $\mathfrak{g}$, and the exponential map from $\mathfrak{g}$ to V is merely the identity. We then have:

(1.2) PROPOSITION. *Let G be a connected and simply connected nilpotent Lie group with Lie algebra $\mathfrak{g}$. Then:*

(a) *The exponential map is a diffeomorphism from $\mathfrak{g}$ to G.*

(b) *If G is identified with $\mathfrak{g}$ via $\exp$, the group law $(x,y) \to xy$ is a polynomial map.*

(c) *If λ denotes Lebesgue measure on $\mathfrak{g}$, then $\lambda \circ \exp^{-1}$ is a bi-invariant Haar measure on G.*

Proof: (a) and (b) are immediate from the preceding remarks, since G is uniquely determined by $\mathfrak{g}$ up to isomorphism. To prove (c), let $\mathfrak{g}_{(1)}, \ldots, \mathfrak{g}_{(m)}, \mathfrak{g}_{(m+1)} = \{0\}$ be the lower central series for $\mathfrak{g}$, and let $n = \dim \mathfrak{g}$ and $n_j = \dim \mathfrak{g}_{(j)}$. Choose a basis $X_{n-n_m+1}, \ldots, X_n$ for $\mathfrak{g}_{(m)}$, extend it to a basis $X_{n-n_{m-1}+1}, \ldots, X_n$ for $\mathfrak{g}_{(m-1)}$, and so forth, obtaining eventually a basis $X_1, \ldots, X_n$ for $\mathfrak{g}$. Let $\xi_1, \ldots, \xi_n$ be the dual basis for $\mathfrak{g}^*$, and let $\eta_k = \xi_k \circ \exp^{-1}$. Thus $\eta_1, \ldots, \eta_n$ are a system of global coordinates on G. By the Campbell-Hausdorff formula and the construction of the η_k's we have

$$\eta_k(xy) = \eta_k(x) + \eta_k(y) + P_k(x,y)$$

where $P_k(x,y)$ depends only on the coordinates $\eta_i(x)$, $\eta_i(y)$ with $i < k$. Therefore, the differentials of the maps $x \to xy$ (y fixed) and $y \to xy$ (x fixed) are given with respect to the coordinates η_k by lower triangular matrices with ones on the diagonal, and their determinants are therefore identically one. It follows that the volume form $d\eta_1 \cdots d\eta_n$ on G, which corresponds to Lebesgue measure on $\mathfrak{g}$, is left and right invariant. #

If $\mathfrak{g}$ is a Lie algebra, a family of <u>dilations</u> on $\mathfrak{g}$ is a family $\{\delta_r : r > 0\}$ of algebra automorphisms of $\mathfrak{g}$ of the form $\delta_r = \exp(A \log r)$ where A is a diagonalizable linear operator on $\mathfrak{g}$ with positive eigenvalues. In particular, $\delta_{rs} = \delta_r\delta_s$ for all $r,s > 0$. If $\alpha > 0$ and $\{\delta_r\}$ is a family of dilations on $\mathfrak{g}$, then so is $\{\tilde{\delta}_r\}$ where $\tilde{\delta}_r = \delta_{r^\alpha} = \exp(\alpha A \log r)$. Hence, by adjusting α if necessary, <u>we shall assume that the smallest eigenvalue of</u> A <u>is</u> 1.

Let $\mathcal{A}$ denote the set of eigenvalues of A, and for $a \in \mathcal{A}$ let $W_a \subset \mathfrak{g}$ be the corresponding eigenspace of A. Thus, $\delta_r X = r^a X$ for $X \in W_a$. If $X \in W_a$ and $Y \in W_b$ then $\delta_r[X,Y] = [\delta_r X, \delta_r Y] = r^{a+b}[X,Y]$ and hence $[W_a,W_b] \subset W_{a+b}$. In particular, since $a \geq 1$ for $a \in \mathcal{A}$, we see that $\mathfrak{g}_{(j)} \subset \oplus_{a \geq j} W_a$. Since the set $\mathcal{A}$ is finite, it follows that $\mathfrak{g}_{(j)} = \{0\}$ for j sufficiently large. Thus:

(1.3) PROPOSITION. _If_ $\mathfrak{g}$ _admits a family of dilations then_ $\mathfrak{g}$ _is nilpotent_.

(On the other hand, not every nilpotent Lie algebra admits a family of dilations: see Dyer [1].)

A Lie algebra $\mathfrak{g}$ is called _graded_ if it is endowed with a vector space decomposition $\mathfrak{g} = \oplus_1^\infty V_k$ (where all but finitely many of the V_k's are $\{0\}$) such that $[V_i, V_j] \subset V_{i+j}$. If $\mathfrak{g}$ is graded, there is a natural family of dilations on $\mathfrak{g}$: namely, if $\alpha = \min\{k : V_k \neq \{0\}\}$ we define $\delta_r = \exp(A \log r)$ where A is the operator defined by $AX = (k/\alpha)X$ for $X \in V_k$. Conversely, if $\mathfrak{g}$ has a family of dilations such that the eigenvalues of A are all rational, then $\mathfrak{g}$ has a natural gradation: if α is the least common denominator of the eigenvalues, we define V_k to be the eigenspace of A with eigenvalue k/α.

A Lie algebra $\mathfrak{g}$ is called _stratified_ if $\mathfrak{g}$ is graded and V_1 generates $\mathfrak{g}$ as an algebra. In this case, if $\mathfrak{g}$ is nilpotent of step m we have

$$\mathfrak{g} = \oplus_1^m V_j, \qquad [V_1, V_j] = V_{j+1},$$

and the natural dilations on $\mathfrak{g}$ are given by

$$\delta_r(\Sigma_1^m X_j) = \Sigma_1^m r^j X_j \qquad (X_j \in V_j).$$

A _homogeneous group_ is a connected and simply connected nilpotent Lie group G whose Lie algebra $\mathfrak{g}$ is endowed with a family of dilations $\{\delta_r\}$. If $\mathfrak{g}$ is graded or stratified, we shall apply the same adjectives to G. If G is a homogeneous group, the maps $\exp \circ \delta_r \circ \exp^{-1}$ are group automorphisms of G; we shall denote them also by δ_r and call them dilations on G. Actually, most of the time we shall write rx instead of $\delta_r x$ for $r > 0$, $x \in G$, reserving the notation δ_r for occasions when it is required for clarity or when we wish to consider δ_r as a mapping. Sometimes we shall even write x/r for $\delta_{1/r}x$. This notation is suggested by the case $G = \mathbb{R}^n$, in which the natural dilations are given by scalar multiplication. (Note, however, that we write the group law multiplicatively, so the distributive law — that is, the automorphism property of δ_r — becomes $r(xy) = (rx)(ry)$.) The analogy with $\mathbb{R}^n$ is strong enough to persuade us to denote the group identity by 0 and to refer to it as the origin. The equation $xx^{-1} = 0$ thus looks peculiar, but the equation $\lim_{r \to 0} rx = 0$ comes out looking right.

Some examples may be in order.

(1) <u>Abelian groups</u>: $\mathbb{R}^n$ is a homogeneous group with dilations given by scalar multiplication.

(2) <u>Heisenberg groups</u>: If n is a positive integer, the Heisenberg group H_n is the group whose underlying manifold is $\mathbb{C}^n \times \mathbb{R}$ and whose multiplication is given by

$$(z_1,\ldots,z_n,t)(z_1',\ldots,z_n',t') = (z_1+z_1',\ldots,z_n+z_n',t+t'+2\,\mathrm{Im}\,\Sigma_1^n z_j \bar{z}_j')$$

H_n is a homogeneous group with dilations

$$\delta_r(z_1,\ldots,z_n,t) = (rz_1,\ldots,rz_n,r^2t).$$

(3) <u>Upper triangular groups</u>: Let G be the group of all $n \times n$ real matrices (a_{ij}) such that $a_{ii} = 1$ for $1 \le i \le n$ and $a_{ij} = 0$ when $i > j$.

G is a homogeneous group with dilations

$$\delta_r(a_{ij}) = (r^{j-i}a_{ij}).$$

These examples are all stratified groups. It is also possible to define other families of dilations on these groups. For instance, on $\mathbb{R}^n$ we can define

$$\delta_r(x_1,\ldots,x_n) = (r^{d_1}x_1,\ldots,r^{d_n}x_n)$$

where $1 = d_1 \le d_2 \le \cdots \le d_n$, and on H_n we can define

$$\delta_r(x_1+iy_1,\ldots,x_n+iy_n,t) = (r^{a_1}x_1+ir^{b_1}y_1,\ldots,r^{a_n}x_n+ir^{b_n}x_n,r^c t)$$

where $\min\{a_1,\ldots a_n,b_1,\ldots,b_n\} = 1$ and $a_j+b_j = c$ for all j. However, when we refer to $\mathbb{R}^n$ or H_n we shall assume that they are equipped with the natural dilations defined in (1) and (2) above unless we state otherwise.

Henceforth we shall be working on a fixed homogeneous group G of dimension n with dilations $\{\delta_r = \exp(A \log r)\}$. We denote by $d_1,\ldots,d_n$ the eigenvalues of A, listed in increasing order and with each eigenvalue listed as many times as its multiplicity, and we set $\bar{d} = \max\{d_j\}$. Thus

$$1 = d_1 \le d_2 \le \cdots \le d_n = \bar{d}.$$

We also fix a basis $X_1,\ldots,X_n$ of $\mathfrak{g}$ such that $AX_j = d_jX_j$ for each j, and we define a Euclidean norm $\|\cdot\|$ on $\mathfrak{g}$ by declaring the X_j's to be orthonormal. We may also regard this norm as a function on G in the obvious way: $\|x\| = \|\exp^{-1}x\|$.

The Euclidean norm is of limited utility for our purposes, since it does not interact in a simple fashion with dilations. We therefore define a <u>homogeneous norm</u> on G to be a continuous function $x \to |x|$ from G to $[0,\infty)$ which is C^∞ on $G\backslash\{0\}$ and satisfies (a) $|x^{-1}| = |x|$ and $|rx| = r|x|$ for all $x \in G$ and $r > 0$, (b) $|x| = 0$ if and only if $x = 0$. Homogeneous norms always exist; one example may be constructed as follows. Observe that

$$X = \Sigma_1^n c_j X_j \in \mathfrak{g} \quad \text{implies} \quad \|\delta_r X\| = (\Sigma_1^n c_j^2 r^{2d_j})^{\frac{1}{2}}.$$

If $X \neq 0$, then, $\|\delta_r X\|$ is a strictly increasing function of r which tends to 0 or ∞ along with r. Hence there is a unique $r(X) > 0$ such that $\|\delta_{r(X)} X\| = 1$, and we may define a homogeneous norm on G by setting $|0| = 0$ and $|x| = 1/r(\exp^{-1} x)$ for $x \neq 0$. (The fact that this function is C^∞ on $G\backslash\{0\}$ follows from the implicit function theorem since the Euclidean unit sphere is a C^∞ manifold.)

<u>Henceforth we assume that G is equipped with a fixed homogeneous norm</u>.

If $x \in G$ and $r > 0$ we define

$$B(r,x) = \{y \in G : |x^{-1}y| < r\},$$

and we call $B(r,x)$ the <u>ball of radius r about x</u>. We observe that $B(r,x)$ is the left translate by x of $B(r,0)$, which in turn is the image under δ_r of $B(1,0)$.

(1.4) LEMMA. <u>For all</u> $x \in G$ <u>and</u> $r > 0$, $\overline{B(r,x)}$ <u>is compact</u>.

Proof: Let us define

$$\rho(x) = \Sigma_1^n |c_j|^{1/d_j} \qquad \text{for} \quad x = \exp(\Sigma_1^n c_j X_j)$$

Then ρ satisfies all the properties of a homogeneous norm except that it is merely continuous instead of C^∞. Clearly $\{x : \rho(x) = 1\}$ is compact and does not contain 0, so the function $x \to |x|$ assumes a positive minimum η on it. Since $|rx| = r|x|$ and $\rho(rx) = r\rho(x)$, it follows that $|x| \geq \eta\rho(x)$ for all x, and hence that $\overline{B(\eta,0)} \subset \{x : \rho(x) \leq 1\}$. Thus $\overline{B(\eta,0)}$ is compact, and it follows by dilation and translation that $\overline{B(r,x)}$ is compact for all $r > 0$, $x \in G$. #

(1.5) PROPOSITION. *There exist* $C_1, C_2 > 0$ *such that*

$$C_1\|x\| \leq |x| \leq C_2\|x\|^{1/\bar{d}} \quad \textit{whenever} \quad |x| \leq 1.$$

Proof: If $y = \exp(\Sigma c_j X_j)$ then $\|ry\| = (\Sigma c_j^2 r^{2d_j})^{1/2}$ and hence $r^{\bar{d}}\|y\| \leq \|ry\| \leq r\|y\|$ for $r \leq 1$. By Lemma 1.4, the Euclidean norm assumes a positive maximum C_1^{-1} and a positive minimum $C_2^{-\bar{d}}$ on $\{y : |y| = 1\}$. Any $x \neq 0$ can be written as $x = |x|y$ where $|y| = 1$, so that for $|x| \leq 1$,

$$|x| \leq |x|\|y\| \leq C_1^{-1}|x|, \qquad \|x\| \geq |x|^{\bar{d}}\|y\| \geq C_2^{-\bar{d}}|x|^{\bar{d}}. \qquad \#$$

(1.6) PROPOSITION. *There is a constant* $C > 0$ *such that for all* $x,y \in G$,

$$|xy| \leq C(|x| + |y|).$$

Proof: By Lemma 1.4, the set $\{(x,y) \in G \times G : |x| + |y| = 1\}$ is compact, so the function $(x,y) \to |xy|$ assumes a finite maximum C on it. Then, given any $x,y \in G$, set $r = |x| + |y|$. It follows that

$$|xy| = r|r^{-1}(xy)| = r|(r^{-1}x)(r^{-1}y)| \leq Cr = C(|x| + |y|). \quad \#$$

By Proposition 1.2, Lebesgue measure on $\mathfrak{g}$ induces a bi-invariant Haar measure on G. We now fix the normalization of Haar measure on G by requiring that the measure of $B(1,0)$ be 1. (Thus, if $G = \mathbb{R}^n$ with the usual norm, our measure is $\Gamma((n+2)/2)/\pi^{n/2}$ times Lebesgue measure.) We shall denote the measure of any measurable $E \subset G$ by $|E|$, and we shall denote the integral of a function f with respect to this measure simply by $\int f$ or $\int f(x)dx$.

The number

$$Q = \Sigma_1^n d_j = \mathrm{trace}(A)$$

will be called the homogeneous dimension of G. Clearly we have

$$|\delta_r(E)| = r^Q|E|, \qquad d(rx) = r^Q dx.$$

In particular, $|B(r,x)| = r^Q$ for all $r > 0$, $x \in G$. Henceforth Q will always denote the homogeneous dimension of G.

A function f on $G\backslash\{0\}$ will be called homogeneous of degree λ $(\lambda \in \mathbb{C})$ if $f \circ \delta_r = r^\lambda f$ for $r > 0$. We note that for any f and g,

$$\int f(x) g \circ \delta_r(x)dx = r^{-Q} \int f \circ \delta_{1/r}(x)g(x)dx$$

provided that the integrals exist. Hence we extend the map $f \to f \circ \delta_r$ to distributions by defining, for any distribution f and any text function ϕ,

$$\langle f \circ \delta_r, \phi \rangle = r^{-Q} \langle f, \phi \circ \delta_{1/r} \rangle ,$$

and we say that a distribution f is _homogeneous of degree_ λ if $f \circ \delta_r = r^\lambda f$. Also, we say that a linear differential operator D on G is _homogeneous of degree_ λ if $D(f \circ \delta_r) = r^\lambda (Df) \circ \delta_r$ for any f. We observe that if D is homogeneous of degree λ and f is homogeneous of degree μ then Df is homogeneous of degree $\mu - \lambda$.

(1.7) PROPOSITION. _Suppose_ f _is of class_ $C^{(1)}$ _on_ $G\backslash\{0\}$ _and homogeneous of degree_ λ $(\lambda \in \mathbb{R})$. _There is a constant_ $C > 0$ _such that_

$$|f(xy) - f(x)| \leq C|y||x|^{\lambda-1} \quad \text{\underline{whenever}} \quad |y| \leq |x|/2.$$

Proof: Both sides of the desired inequality are homogeneous of degree λ, so it suffices to assume that $|x| = 1$ and $|y| \leq 1/2$. In this case x and xy are bounded and bounded away from zero, and the map $y \to xy$ is smooth, so by the mean value theorem and Proposition 1.5,

$$|f(xy) - f(x)| \leq C\|y\| \leq C'|y| = C'|y||x|^{\lambda-1}. \quad \#$$

This proposition may be applied, in particular, to the homogeneous norm. Combining this with Proposition 1.6, we see that there exists a constant γ such that

(1.8) $$|xy| \leq \gamma\,(|x| + |y|) \qquad \text{for all } x,y \in G,$$

(1.9) $\big||xy| - |x|\big| \leq \gamma|y|$ for all $x,y \in G$ such that $|y| \leq |x|/2$.

Henceforth γ will always denote the minimal constant satisfying (1.8) and (1.9). Clearly $\gamma \geq 1$. We shall use (1.8) and (1.9) without comment in the sequel.

The following simple fact will be useful later:

(1.10) LEMMA. For all $x,y \in G$ and $s > 0$,

$$(1+|x|)^s(1+|y|)^{-s} \leq \gamma^s(1+|xy^{-1}|)^s.$$

Proof: Since $|x| \leq \gamma(|xy^{-1}|+|y|)$ we have

$$1 + |x| \leq \gamma(1+|xy^{-1}|)(1+|y|),$$

and we obtain the desired inequality by raising both sides to the s-th power. #

We now establish our notation for some common function spaces on G (with apologies for some slight inconsistencies). If $\Omega \subset G$, $C(\Omega)$, $(C_c(\Omega), C_0(\Omega))$ denotes the space of continuous functions on G (with compact support, vanishing at infinity). If Ω is open, $C^{(k)}(\Omega)$ denotes the class of k times continuously differentiable functions on Ω, $C^\infty(\Omega) = \bigcap_1^\infty C^{(k)}(\Omega)$, and $C_0^\infty(\Omega) = C^\infty(\Omega) \cap C_c(\Omega)$. When $\Omega = G$ we shall usually omit mentioning it. If $0 < p \leq \infty$, L^p will denote the usual Lebesgue space on G. For $0 < p < \infty$ we write

$$\|f\|_p = \left(\int |f(x)|^p dx\right)^{1/p},$$

even though this is not a norm for $p < 1$. (The map $(f,g) \to \|f-g\|_p^p$ is, however, a metric on L^p for $p < 1$.)

We recall that if f is a measurable function on G, its <u>distribution function</u> $\lambda_f : [0,\infty) \to [0,\infty]$ is defined by

$$(1.11) \qquad \lambda_f(\alpha) = |\{x : |f(x)| > \alpha\}|,$$

and its <u>nonincreasing rearrangement</u> $f^* : [0,\infty) \to [0,\infty]$ is defined by

$$(1.12) \qquad f^*(t) = \inf\{\alpha : \lambda_f(\alpha) \leq t\}.$$

As is well known (cf. Stein and Weiss [2]),

$$\int_G |f(x)|^p dx = -\int_0^\infty \alpha^p d\lambda_f(\alpha) = p\int_0^\infty \alpha^{p-1}\lambda_f(\alpha)d\alpha = \int_0^\infty f^*(t)^p dt.$$

For $0 < p < \infty$, <u>weak</u> L^p is the space of functions f such that

$$[f]_p = \sup_{\alpha>0} \alpha^p\lambda_f(\alpha) = \sup_{t>0} t^{1/p} f^*(t) < \infty.$$

It is easily checked that $[f]_p \leq \|f\|_p$ (Chebyshev's inequality). $[\ ]_p$ is not a norm, but it defines a topology on weak L^p. A subadditive operator which is bounded from L^p to weak L^q is said to be of <u>weak type</u> (p,q).

We conclude this section with some results concerning integration in "polar coordinates".

(1.13) PROPOSITION. *If f is continuous on $G\backslash\{0\}$ and homogeneous of degree $-Q$, there is a constant μ_f (the "mean value" of f) such that for all $g \in L^1((0,\infty), r^{-1}dr)$,*

$$(1.14) \qquad \int_G f(x)g(|x|)dx = \mu_f \int_0^\infty g(r)r^{-1}dr.$$

<u>Proof</u>: Define $L_f : (0,\infty) \to \mathbb{C}$ by

$$L_f(r) = \int_{1\leq|x|\leq r} f(x)dx \quad \text{if } r \geq 1,$$

$$= -\int_{r\leq|x|\leq 1} f(x)dx \quad \text{if } r < 1.$$

By making the change of variable $x \to sx$ and using the homogeneity of f, one easily checks that $L_f(rs) = L_f(r) + L_f(s)$ for all $r,s > 0$. Since L is continuous, it follows that $L_f(r) = L_f(e)\log r$, and we set $\mu_f = L_f(e)$. Then equation (1.14) is obvious when g is the characteristic function of an interval, and it follows in general by taking linear combinations and limits of such functions. #

(1.15) PROPOSITION. <u>Let</u> $S = \{x \in G : |x| = 1\}$. <u>There is a (unique) Radon measure</u> σ <u>on</u> S <u>such that for all</u> $u \in L^1(G)$,

$$\int_G u(x)dx = \int_0^\infty \int_S u(ry)r^{Q-1}d\sigma(y)dr.$$

<u>Proof</u>: If $f \in C(S)$, let us define $\tilde{f}$ on $G\setminus\{0\}$ by $\tilde{f}(x) = |x|^{-Q}f(|x|^{-1}x)$: then $\tilde{f}$ satisfies the hypotheses of Proposition 1.13. The map $f \to \mu_{\tilde{f}}$ is clearly a positive linear functional on $C(S)$, so it is given by integration against a Radon measure σ on S. If $g \in C_c((0,\infty))$, then, we have

$$\int_G f(|x|^{-1}x)g(|x|)dx = \int_G \tilde{f}(x)|x|^Q g(|x|)dx = \mu_{\tilde{f}} \int_0^\infty r^{Q-1}g(r)dr$$

$$= \int_0^\infty \int_S f(y)g(r)r^{Q-1}d\sigma(y)dr.$$

Since linear combinations of functions of the form $f(|x|^{-1}x)g(|x|)$ are dense in $L^1(G)$, the theorem follows. #

Remark: The measure σ can be shown to have a smooth density: cf. Fabes and Rivière [1] and Goodman [2].

(1.16) COROLLARY. _Let_ $C = \sigma(S)$. _Then if_ $0 < a < b < \infty$ _and_ $\alpha \in \mathbb{C}$,

$$\int_{a<|x|<b} |x|^{\alpha-Q}dx = C\alpha^{-1}(b^\alpha - a^\alpha) \qquad \text{if } \alpha \neq 0,$$

$$= C\log(b/a) \qquad \text{if } \alpha = 0.$$

(1.17) COROLLARY. _Suppose_ $\alpha \in \mathbb{R}$ _and_ f _is a measurable function on_ G _such that_ $|f(x)| = O(|x|^{\alpha-Q})$. _If_ $\alpha > 0$ _then_ f _is integrable near_ 0, _and if_ $\alpha < 0$ _then_ f _is integrable near_ ∞.

These two corollaries will be frequently used without comment in the sequel.

B. Convolutions

If f and g are measurable functions on G, their _convolution_ $f*g$ is defined by

$$f*g(x) = \int f(y)g(y^{-1}x)dy = \int f(xy^{-1})g(y)dy,$$

provided that the integrals converge. The following two propositions give the basic facts about convolution of L^p and weak L^p functions. Implicit in each of them is the fact that, under the stated hypotheses, the integral defining $f*g(x)$ converges absolutely for almost every x, as may be seen by an examination of the proofs.

(1.18) PROPOSITION. *Suppose* $1 \leq p,q,r \leq \infty$ *and* $p^{-1}+q^{-1} = r^{-1}+1$. *If* $f \in L^p$ *and* $g \in L^q$ *then* $f*g \in L^r$ *and*

$$\|f*g\|_r \leq \|f\|_p \|g\|_q \qquad \text{(\underline{Young's inequality})}.$$

Proof: Case I: $r = \infty$. Then p and q are conjugate exponents and the result is immediate from Hölder's inequality.

Case II: $r = q$, $p = 1$. Let q' be the conjugate exponent to q. By Hölder's inequality,

$$|f*g(x)| \leq \int |f(xy^{-1})|^{(1/q)+(1/q')}|g(y)|dy$$

$$\leq \left(\int |f(xy^{-1})|dy\right)^{1/q'}\left(\int |f(xy^{-1})||g(y)|^q dy\right)^{1/q}$$

$$= \|f\|_1^{1/q'}\left(\int |f(xy^{-1})||g(y)|^q dy\right)^{1/q}.$$

Thus by Fubini's theorem,

$$\int |f*g(x)|^q dx \leq \|f\|_1^{q/q'} \iint |f(xy^{-1})||g(y)|^q dydx$$

$$= \|f\|_1^{(q/q')+1}\|g\|_q^q,$$

so that $\|f*g\|_q \leq \|f\|_1\|g\|_q$.

The general case follows from Case I, Case II, and the Riesz-Thorin convexity theorem (cf. Stein and Weiss [2]). #

(1.19) PROPOSITION. *Suppose* $1 \leq p < \infty$, $1 < q,r < \infty$, *and* $p^{-1} + q^{-1} = r^{-1} + 1$. *If* $f \in L^p$ *and* $g \in \text{weak } L^q$ *then* $f*g \in \text{weak } L^r$, *and there exists* $C_1 = C_1(p,q)$ *such that* $[f*g]_r \leq C_1 \|f\|_p [g]_q$. *Moreover, if* $p > 1$ *then* $f*g \in L^r$ *and there exists* $C_2 = C_2(p,q)$ *such that* $\|f*g\|_r \leq C_2 \|f\|_p [g]_q$.

Proof: We first observe that the strong result for $p > 1$ follows from the weak result by the Marcinkiewicz interpolation theorem (cf. Stein and Weiss [2]). Suppose then that $f \in L^p$ and $g \in \text{weak } L^q$, and (without loss of generality) that $\|f\|_p = [g]_q = 1$. Given $\alpha > 0$, set $M = (\alpha/2)^{r/q}(q/r)^{r/qp'}$ where p' is the conjugate exponent to p. Define $g_1(x) = g(x)$ if $|g(x)| \leq M$ and $g_1(x) = 0$ otherwise, and set $g_2 = g - g_1$. Since

$$\lambda_{f*g}(\alpha) \leq \lambda_{f*g_1}(\alpha/2) + \lambda_{f*g_2}(\alpha/2),$$

it suffices to show that each term on the right is bounded by $C\alpha^{-r}$ where C depends only on p and q. On the one hand, since $q^{-1} - (p')^{-1} = r^{-1} > 0$, we have $p' - q > 0$ and hence

$$\int |g_1(x)|^{p'} dx = p'\int_0^\infty \alpha^{p'-1}\lambda_{g_1}(\alpha)d\alpha \leq p'\int_0^M \alpha^{p'-1}\lambda_g(\alpha)d\alpha$$

$$\leq p'\int_0^M \alpha^{p'-1-q}d\alpha = \frac{p'}{p'-q}M^{p'-q} = \frac{r}{q}M^{qp'/r} = (\alpha/2)^{p'}.$$

Thus for all $x \in G$,

$$|f*g_1(x)| \leq \|f\|_p \|g_1\|_{p'} \leq \alpha/2,$$

which implies that $\lambda_{f*g_1}(\alpha/2) = 0$. On the other hand, since $q > 1$,

$$\int |g_2(x)|dx = \int_0^\infty \lambda_{g_2}(\alpha)d\alpha = \int_0^M \lambda_g(M)d\alpha + \int_M^\infty \lambda_g(\alpha)d\alpha$$

$$\leq M\cdot M^{-q} + \int_M^\infty \alpha^{-q}d\alpha = \frac{q}{q-1}M^{1-q},$$

and hence by Proposition 1.18,

$$\|f*g_2\|_p \leq \|f\|_p\|g_2\|_1 \leq q(q-1)^{-1}M^{1-q}.$$

But then

$$\lambda_{f*g_2}(\alpha/2) \leq [2\|f*g_2\|_p/\alpha]^p$$

$$\leq (\frac{2}{\alpha})^p(\frac{q}{q-1})^p(\frac{\alpha}{2})^{pr(1-q)/q}(\frac{q}{r})^{pr(1-q)/qp'}$$

$$= C(p,q)\alpha^{-r},$$

so we are done. #

We now summarize the basic facts about approximations to the identity. The following notation will be used throughout this monograph: if ϕ is a function on G and $t > 0$, we define ϕ_t by

$$\phi_t = t^{-Q}\phi \circ \delta_{1/t}, \quad \text{i.e.,} \quad \phi_t(x) = t^{-Q}\phi(x/t).$$

We observe that if $\phi \in L^1$ then $\int \phi_t(x)dx$ is independent of t.

(1.20) PROPOSITION. *Suppose* $\phi \in L^1$ *and* $\int \phi(x)dx = a$. *Then*:

(i) *If* $f \in L^p$ $(1 \leq p < \infty)$, *then* $\|f*\phi_t - af\|_p \to 0$ as $t \to 0$.

(ii) *If* f *is bounded and right uniformly continuous, then* $\|f*\phi_t - af\|_\infty \to 0$ *as* $t \to 0$.

(iii) *If* f *is bounded on* G *and continuous on an open set* $\Omega \subset G$, *then* $f*\phi_t - af \to 0$ *uniformly on compact subsets of* Ω as $t \to 0$.

Proof: If f is any function on G and $y \in G$, let us define $f^y(x) = f(xy^{-1})$. If $f \in L^p$, $1 \leq p < \infty$, it is easily seen that

(1.21) $$\|f^y - f\|_p \to 0 \quad \text{as} \quad y \to 0.$$

(Use the fact that C_c is dense in L^p.) If $p = \infty$, (1.21) holds if and only if f is (almost everywhere equal to) a right uniformly continuous function. We observe that

$$\begin{aligned} f*\phi_t(x) - af(x) &= \int f(xy^{-1})t^{-Q}\phi(y/t)dy - af(x) \\ &= \int f(x(tz)^{-1})\phi(z)dz - af(x) \\ &= \int [f(x(tz)^{-1}) - f(x)]\phi(z)dz. \end{aligned}$$

Hence by Minkowski's inequality,

$$\|f*\phi_t - af\|_p \leq \int \|f^{tz} - f\|_p |\phi(z)| dz.$$

Since $\|f^{tz} - f\|_p \leq 2\|f\|_p$, under the hypothesis of (i) or (ii) it follows from (1.21) and the dominated convergence theorem that $\|f*\phi_t - af\|_p \to 0$.

The routine modification of this argument (with $p = \infty$) needed to establish (iii) is left to the reader. #

C. Derivatives and Polynomials

There are three common ways of viewing the elements of the Lie algebra $\mathfrak{g}$ of G: (1) as tangent vectors at the origin, (2) as left-invariant vector fields, (3) as right-invariant vector fields. We shall have no use for the first interpretation, but we shall need both of the others. Accordingly, let us denote by $\mathfrak{g}_L$ and $\mathfrak{g}_R$ the spaces of left-invariant and right-invariant vector fields on G. We shall follow the usual custom of identifying $\mathfrak{g}$ with $\mathfrak{g}_L$, and in particular we shall think of the exponential map as going from $\mathfrak{g}_L$ to G. (This point is not entirely trivial, since the map which sends $X \in \mathfrak{g}_L$ to the unique $\tilde{X} \in \mathfrak{g}_R$ which agrees with X at the origin is an anti-isomorphism rather than an isomorphism: that is, $[X,Y]^{\sim} = [\tilde{Y},\tilde{X}]$. Hence if one applies the Campbell-Hausdorff formula to $\mathfrak{g}_R$, one obtains a different, although isomorphic, group law for G.)

We recall that in Section A we fixed a basis $X_1,\ldots,X_n$ for $\mathfrak{g}$ consisting of eigenvectors for the dilations δ_r with eigenvalues $r^{d_1},\ldots,r^{d_n}$. The X_j's are now to be regarded as left-invariant differential operators on G, and we denote by $Y_1,\ldots,Y_n$ the corresponding basis for $\mathfrak{g}_R$: that is, Y_j is the element of $\mathfrak{g}_R$ such that $Y_j|_0 = X_j|_0$. Thus, for $f \in C^{(1)}$,

$$X_j f(y) = \frac{d}{dt} f(y\cdot\exp(tX_j))\big|_{t=0}, \qquad Y_j f(y) = \frac{d}{dt} f(\exp(tX_j)\cdot y)\big|_{t=0}.$$

The differential operators X_j and Y_j are homogeneous of degree d_j, for

$$X_j(f \circ \delta_r)(y) = \frac{d}{dt} f((ry)\exp(r^{d_j} tX_j))|_{t=0}$$

$$= r^{d_j} \frac{d}{dt} f((ry)\exp(tX_j))|_{t=0} = r^{d_j}(X_j f \circ \delta_r)(y),$$

and similarly for Y_j.

We adopt the following multiindex notation for higher order derivatives. If $I = (i_1,\ldots,i_n) \in \mathbb{N}^n$, we set

$$X^I = X_1^{i_1} X_2^{i_2} \cdots X_n^{i_n}, \qquad Y^I = Y_1^{i_1} Y_2^{i_2} \cdots Y_n^{i_n}.$$

By the Poincaré-Birkhoff-Witt theorem (cf. Bourbaki [1], I.2.7), the operators X^I(resp. Y^I) form a basis for the algebra of left-invariant (resp. right-invariant) differential operators on G. Further, we set

$$|I| = i_1+i_2+\cdots+i_n, \qquad d(I) = d_1 i_1+d_2 i_2+\cdots+d_n i_n.$$

Thus $|I|$ is the order of the differential operators X^I and Y^I, while $d(I)$ is their degree of homogeneity, or, as we shall say, their <u>homogeneous degree</u>. We shall denote by Δ the additive sub-semigroup of $\mathbb{R}$ generated by $0,d_1,\ldots,d_n$. In other words, Δ is the set of all numbers $d(I)$ as I ranges over $\mathbb{N}^n$. We observe that $\mathbb{N} \subset \Delta$ since $d_1 = 1$.

We pause to make two useful remarks. First, since left and right translations are isometries on L^2, the operators X_j and Y_j are formally skew-adjoint. Thus,

$$\int (X^I f)g = (-1)^{|I|}\int f(X^I g), \qquad \int f(Y^I g) = (-1)^{|I|}\int (Y^I f)g,$$

for all smooth f and g such that the integrands decay suitably at infinity. Second, the operators X^I and Y^I interact with convolutions in the following way:

$$X^I(f*g) = f*(X^I g), \quad Y^I(f*g) = (Y^I f)*g, \quad (X^I f)*g = f*(Y^I g).$$

The first two of these equations are established by differentiating under the integral sign, while the third is established by integration by parts:

$$(X^I f)*g(x) = \int X^I f(xy)g(y^{-1})dy = (-1)^{|I|}\int f(xy)X^I[g(y^{-1})]dy$$

$$= \int f(xy)(Y^I g)(y^{-1})dy = f*(Y^I g)(x).$$

We now investigate polynomials on G. A function P on G will be called a <u>polynomial</u> if $P \circ \exp$ is a polynomial on $\mathfrak{g}$. We denote by $\xi_1,\ldots,\xi_n$ the basis for the linear forms on $\mathfrak{g}$ dual to the basis $X_1,\ldots,X_n$ for $\mathfrak{g}$, and we set $\eta_j = \xi_j \circ \exp^{-1}$. Then $\eta_1,\ldots,\eta_n$ are polynomials on G which form a global coordinate system on G and generate the algebra of polynomials on G. Thus, every polynomial on G can be written uniquely as

$$P = \Sigma_I a_I \eta^I \quad (\eta^I = \eta_1^{i_1} \cdots \eta_n^{i_n},\ a_I \in \mathbb{C})$$

where all but finitely many of the coefficients a_I vanish. Clearly η^I is homogeneous of degree $d(I)$, so the set of possible degrees of homogeneity for polynomials is the set Δ introduced above. If $P = \Sigma a_I \eta^I$, we shall refer to its degree — that is, $\max\{|I| : a_I \neq 0\}$ — as its <u>isotropic degree</u>. Further, we define its <u>homogeneous degree</u> to be $\max\{d(I) : a_I \neq 0\}$. If $N \in \mathbb{N}$ we denote the space of polynomials of isotropic degree $\leq N$ by

P_N^{iso}, and if $a \in \Delta$ we denote the space of polynomials of homogeneous degree $\leq a$ by P_a. Since $1 \leq d_j \leq \bar{d}$ for $j = 1,\dots,n$, we clearly have $P_N \subset P_N^{iso} \subset P_{\bar{d}N}$ for $N \in \mathbb{N}$.

We can now give a more explicit description of the group law in terms of the coordinates η_j. Since the map $(x,y) \to \eta_j(xy)$ is a polynomial on $G \times G$ which is jointly homogeneous of degree d_j (that is, $\eta_j((rx)(ry)) = r^{d_j}\eta_j(xy))$, and since the Campbell-Hausdorff formula implies that $\eta_j(xy) = \eta_j(x) + \eta_j(y)$ modulo terms of isotropic degree ≥ 2, we must have

$$(1.22) \qquad \eta_j(xy) = \eta_j(x) + \eta_j(y) + \sum_{I \neq 0, J \neq 0, d(I)+d(J)=d_j} C_j^{IJ} \eta^I(x)\eta^J(y)$$

for some constants C_j^{IJ}. Since the multiindices I and J in (1.22) must satisfy $d(I) < d_j$ and $d(J) < d_j$, it follows that the monomials η^I, η^J can only involve coordinates with degrees of homogeneity less than d_j, and in particular can only involve the coordinates $\eta_1,\dots,\eta_{j-1}$. We note two special cases:

$$(1.23) \qquad d_j = 1: \quad \eta_j(xy) = \eta_j(x) + \eta_j(y).$$

$$(1.24) \qquad d_j = 2: \quad \eta_j(xy) = \eta_j(x) + \eta_j(y) + \sum_{d_k = d_\ell = 1} C_j^{k\ell} \eta_k(x)\eta_\ell(y).$$

(1.25) PROPOSITION. <u>For any</u> $a \in \Delta$, <u>the space</u> P_a <u>is invariant under left and right translations</u>.

<u>Proof</u>: From (1.22) it is clear that $\eta_j(xy)$ is in P_{d_j} as a function of x for each y, and also as a function of y for each x. Since the η_j's generate all polynomials, the result follows immediately. #

Remark: P_N^{iso} is not invariant under translations (unless $N = 0$ or G is Abelian). Consequently, it will not be of much use to us.

(1.26) PROPOSITION. We have

$$X_j = \Sigma P_{jk}(\partial/\partial\eta_k), \qquad Y_j = \Sigma Q_{jk}(\partial/\partial\eta_k), \tag{1.27}$$

where $P_{jj} = Q_{jj} = 1$, $P_{jk} = Q_{jk} = 0$ if $d_k < d_j$ or if $d_k = d_j$ and $k \neq j$, and P_{jk}, Q_{jk} are homogeneous polynomials of degreee $d_k - d_j$ if $d_k > d_j$.

Proof: For $x \in G$ define $L_x : G \to G$ by $L_x(y) = xy$. Then for any differentiable function f on G and $x \in G$,

$$X_j f(x) = (X_j f) \circ L_x(0) = X_j(f \circ L_x)(0) = (\partial/\partial\eta_j)(f \circ L_x)(0)$$

since X_j agrees with $\partial/\partial\eta_j$ at 0. Therefore, by the chain rule,

$$X_j f(x) = \sum_{k=1}^{n} \frac{\partial f}{\partial\eta_k}(x) \frac{\partial[\eta_k \circ L_x]}{\partial\eta_j}(0).$$

(The notation is a bit awkward, but the meaning should be clear.) But by formula (1.22),

$$\frac{\partial[\eta_k \circ L_x]}{\partial\eta_j}(0) = \delta_{jk} + \sum_{d(I)=d_k-d_j} C_k^{I[j]} \eta^I(x)$$

where $[j]$ is the multiindex with 1 in the j-th place and zeros elsewhere. The assertion for X_j now follows immediately, and the assertion for Y_j is proved in the same way. #

There are formulas of exactly the same sort expressing $\partial/\partial\zeta_j$ in terms of the X_k's or Y_k's:

$$\partial/\partial\eta_j = \Sigma P'_{jk}X_k = \Sigma Q'_{jk}Y_k$$

where P'_{jk}, Q'_{jk} are of the same form as P_{jk}, Q_{jk} in (1.27). Indeed, these formulas may be easily obtained from (1.27) by starting with $j = n$ and working backwards: we have

$$X_n = \partial/\partial\eta_n,$$

$$X_{n-1} = \partial/\partial\eta_{n-1} + P_{(n-1)n}\partial/\partial\eta_n,$$

$$X_{n-2} = \partial/\partial\eta_{n-2} + P_{(n-2)(n-1)}\partial/\partial\eta_{n-1} + P_{(n-2)n}\partial/\partial\eta_n,$$

so that

$$\partial/\partial\eta_n = X_n,$$

$$\partial/\partial\eta_{n-1} = X_{n-1} - P_{(n-1)n}X_n,$$

$$\partial/\partial\eta_{n-2} = X_{n-2} - P_{(n-2)(n-1)}(X_{n-1} - P_{(n-1)n}X_n) - P_{(n-2)n}X_n,$$

and so forth.

These formulas yield analogous results for higher order derivatives. For instance, we have

$$X^I = \sum_{|J|\le|I|,d(J)\ge d(I)} P_{IJ}(\partial/\partial\eta)^J \tag{1.28}$$

where P_{IJ} is a homogeneous polynomial of degree $d(J)-d(I)$. Similar formulas yield Y^I in terms of the $(\partial/\partial\eta)^J$'s and $(\partial/\partial\eta)^I$ in terms of the X^J's or Y^J's. Combining these formulas, one immediately deduces the following result.

(1.29) PROPOSITION. *We have*

$$X^I = \sum_{|J|\leq|I|,d(J)\geq d(I)} P_{IJ}Y^J,$$

$$Y^I = \sum_{|J|\leq|I|,d(J)\geq d(I)} Q_{IJ}X^J,$$

where P_{IJ} *and* Q_{IJ} *are homogeneous polynomials of degree* $d(J)-d(I)$.

(1.30) PROPOSITION. *Suppose* $a \in \Delta$, *and let* $\mu = \dim \mathcal{P}_a$. *Then the maps* (i) $P \to ((\partial/\partial\eta)^I P(0))_{d(I)\leq a}$, (ii) $P \to (X^I P(0))_{d(I)\leq a}$, (iii) $P \to (Y^I P(0))_{d(I)\leq a}$ *are linear isomorphisms from* $\mathcal{P}_a$ to $\mathbb{C}^\mu$.

Proof: In case (i) this is a simple consequence of Taylor's theorem. Also, with reference to (1.28), since P_{IJ} is a constant function when $d(I) = d(J)$ and $P_{IJ}(0) = 0$ when $d(J) > d(I)$, we have

$$X^I|_0 = \sum_{|J|\leq|I|,d(J)=d(I)} P_{IJ}(\partial/\partial\eta)^J|_0,$$

and similarly for the other formulas relating X^I, Y^I, and $(\partial/\partial\eta)^I$. Cases (ii) and (iii) follow easily from this observation together with case (i). #

In view of this result, we make the following definition. Suppose $x \in G$, $a \in \Delta$, and f is a function whose (distribution) derivatives $X^I f$ (resp. $Y^I f$) are continuous functions in a neighborhood of x for $d(I) \leq a$. The *left* (resp. *right*) *Taylor polynomial of* f *at* x *of*

<u>homogeneous degree</u> a is the unique $P \in \mathcal{P}_a$ such that $X^I P(0) = X^I f(x)$ (resp. $Y^I P(0) = Y^I f(x)$) for $d(I) \leq a$. We now work toward establishing a version of Taylor's theorem with remainder using these polynomials.

(1.31) LEMMA. <u>The map</u> $\Phi : \mathbb{R}^n \to G$ <u>defined by</u>

$$\Phi(t_1,\ldots,t_n) = \exp(t_1X_1)\exp(t_2X_2) \cdots \exp(t_nX_n)$$

<u>is a global diffeomorphism. Moreover there is a constant</u> C_0 <u>such that</u>

$$|t_j|^{1/d_j} \leq C_0|\Phi(t_1,\ldots,t_n)| \quad \underline{\text{for}} \quad (t_1,\ldots,t_n) \in \mathbb{R}^n, \quad 1 \leq j \leq n.$$

<u>Proof</u>: Φ is clearly a C^∞ map, and the Campbell-Hausdorff formula shows that the differential $d\Phi(0) : \mathbb{R}^n \to T_0G$ is the isomorphism

$$d\Phi(0)(t_1,\ldots,t_n) = \Sigma t_jX_j|_0,$$

so that Φ is a local diffeomorphism near 0. More precisely, there exist δ, $C > 0$ such that Φ is a diffeomorphism from $U = \Phi^{-1}(B(\delta,0))$ to $B(\delta,0)$ and $U \subset \{(t_1,\ldots,t_n) : \max|t_j|^{1/d_j} \leq C\}$. However, for any $r > 0$,

$$\Phi(r^{d_1}t_1,\ldots,r^{d_n}t_n) = r\Phi(t_1,\ldots,t_n). \tag{1.32}$$

Thus if $\Phi(t_1,\ldots,t_n) = \Phi(s_1,\ldots,s_n)$ we have, for all $r > 0$,

$$\Phi(r^{d_1}t_1,\ldots,r^{d_n}t_n) = \Phi(r^{d_1}s_1,\ldots,r^{d_n}s_n).$$

But this is impossible for r sufficiently small unless $t_j = s_j$ for all j, so Φ is injective. Moreover, any $x \in G$ can be written as $x = ry$ with $r > |x|/\delta$ and $y = r^{-1}x \in B(\delta,0)$. It then follows from (1.32) that Φ is surjective and that $|t_j|^{1/d_j} \leq C_0|\Phi(t_1,\ldots,t_n)|$ where $C_0 = C/\delta$. #

(1.33) MEAN VALUE THEOREM. <u>There exist</u> $\overline{C} > 0$ <u>and</u> $\beta > 0$ <u>such that for all</u> f <u>of class</u> $C^{(1)}$ <u>on</u> G <u>and all</u> $x,y \in G$,

$$|f(yx)-f(x)| \leq \overline{C}\Sigma_1^n |y|^{d_j} \sup_{|z|\leq\beta|y|} |Y_j f(zx)|.$$

<u>Proof</u>: First suppose $y = \exp(tX_j)$. Then $|y| = |t|^{1/d_j}|\exp X_j|$, so if we set $C = \max\{|\exp X_k|^{-d_k} : 1 \leq k \leq n\}$ we have

$$(1.34) \qquad |f(xy)-f(x)| = \left|\int_0^t Y_j f(\exp(sX_j)\cdot x)ds\right|$$

$$\leq |t| \sup_{|z|\leq|y|} |Y_j f(zx)|$$

$$\leq C|y|^{d_j} \sup_{|z|\leq|y|} |Y_j f(zx)|.$$

For the general case, we note that by Lemma 1.31, any $y \in G$ can be written uniquely as $y = y_1 y_2 \cdots y_n$ where $y_j = \exp(t_j X_j)$ and

$$(1.35) \qquad |y_j| = |t_j|^{1/d_j} |\exp X_j| \leq C_0 C' |y|$$

where C_0 is as in Lemma 1.31 and $C' = \max\{|\exp X_k| : 1 \leq k \leq n\}$. Thus by (1.34),

$$|f(xy)-f(x)| \leq \Sigma_1^n |f(y_j y_{j+1} \cdots y_n x) - f(y_{j+1} \cdots y_n x)|$$

$$\leq C\Sigma_1^n |y_j|^{d_j} \sup_{|z|\leq|y_j|} |Y_j f(zy_{j+1} \cdots y_n x)|.$$

But if $|z| \leq |y_j|$, repeated application of the triangle inequality (1.8), together with (1.35), yields

$$|zy_{j+1}\cdots y_n| \le \gamma^{j-1}(|y_j|+\cdots+|y_n|) \le n\gamma^{n-1}C_0C'|y|,$$

and therefore, if we set $\beta = n\gamma^{n-1}C_0C'$,

$$|f(yx)-f(x)| \le C\Sigma_1^n(C_0C'|y|)^{d_j}\ \sup_{|z|\le\beta|y|}\ |Y_jf(zx)|$$

$$\le \bar{C}\Sigma_1^n|y|^{d_j}\ \sup_{|z|\le\beta|y|}\ |Y_jf(zx)|.\quad \#$$

The constant β in the mean value theorem — or, rather, the _minimal_ β which makes the mean value theorem true — will occur frequently in the sequel, and _henceforth_ β _will always denote this constant_. We observe that $\beta \ge 1$, with equality when $G = \mathbb{R}^n$ and $|\cdot|$ is the Euclidean norm.

(1.36) LEMMA. _If_ $a \in \Delta$ _then_ $\max\{|I| : d(I) \le a\} = [a]$.

Proof: Since $|I| \le d(I)$ for all I, $d(I) \le a$ implies $|I| \le [a]$. On the other hand, since $d_1 = 1$, for $I = ([a],0,0,\ldots,0)$ we have $|I| = d(I) = [a]$. #

(1.37) THEOREM (TAYLOR INEQUALITY). _Suppose_ $a \in \Delta$ $(a > 0)$, _and_ $k = [a]$. _There is a constant_ $C_a > 0$ _such that for all functions_ f _of class_ $C^{(k+1)}$ _on_ G _and all_ $x,y \in G$,

$$|f(yx)-P_x(y)| \le C_a \sum_{|I|\le k+1,\,d(I)>a} |y|^{d(I)} \sup_{|z|\le\beta^{k+1}|y|} |Y^If(zx)|,$$

where P_x _is the right Taylor polynomial of_ f _at_ x _of homogeneous degree_ a.

<u>Proof</u>: Let $g(y) = f(yx)-P_x(y)$. Then $Y^J g(0) = 0$ for $d(J) \leq a$, while $Y^J g(y) = Y^J f(yx)$ for $d(J) > a$. We shall show by induction on m, $1 \leq m \leq k+1$, that if $a-m < d(J) \leq a$ then

$$(1.38) \qquad |Y^J g(y)| \leq C_m \sum_{|I| \leq k+1, d(I)>a} |y|^{d(I)-d(J)} \sup_{|z| \leq \beta^m |y|} |Y^I f(zx)|.$$

The desired result then follows by taking $m = k+1$ and $J = 0$.

First suppose $a-1 < d(J) \leq a$. Then by the mean value theorem (1.33), since $Y^J g(0) = 0$,

$$|Y^J g(y)| \leq C \sum_{j=1}^n |y|^{d_j} \sup_{|z| \leq \beta|y|} |Y_j Y^J g(z)|.$$

Now $Y_j Y^J$ is a right-invariant differential operator of order $|J|+1$ and homogeneous degree $d(J)+d_j$, so it is a linear combination of the monomials Y^I such that $|I| \leq |J|+1$ and $d(I) = d(J)+d_j$. Moreover, $|J|+1 \leq k+1$ by Lemma 1.36, and $d(J)+d_j \geq d(J)+1 > a$, so that $Y^I g(z) = Y^I f(zx)$. Therefore,

$$|Y^J g(y)| \leq C' \sum_{|I| \leq k+1, d(I)>a} |y|^{d(I)-d(J)} \sup_{|z| \leq \beta|y|} |Y^I f(zx)|.$$

Now suppose (1.38) is true when $a-m+1 < d(J) \leq a$, and suppose $a-m < d(J) \leq a$. By the same reasoning as above we obtain

$$|Y^J g(y)| \leq C' \Sigma_1^n S_j, \quad \text{where}$$

$$(1.39) \qquad S_j = \sum_{|I| \leq |J|+1, d(I)=d(J)+d_j} |y|^{d(I)-d(J)} \sup_{|z| \leq \beta|y|} |Y^I g(z)|.$$

If $d(J)+d_j > a$ then $Y^I g(z) = Y^I f(zx)$, so S_j is dominated by the right hand side of (1.38) (since $\beta \geq 1$). If $d(J)+d_j \leq a$ we have $d(J)+d_j > a-m+1$ and we can apply the inductive hypothesis to $Y^I g(z)$, obtaining

$$\sup_{|z|\leq\beta|y|} |Y^I g(z)|$$

$$\leq \sup_{|z|\leq\beta|y|} C_{m-1} \sum_{|K|\leq k+1, d(K)>a} |z|^{d(K)-d(I)} \sup_{|w|\leq\beta^{m-1}|z|} |Y^K f(wx)|$$

$$\leq C_{m-1} \sum_{|K|\leq k+1, d(K)>a} (\beta|y|)^{d(K)-d(I)} \sup_{|w|<\beta^m|y|} |Y^K f(wx)|.$$

Substituting this into (1.39) we see that S_j is again dominated by the right hand side of (1.38), so we are done. #

Remark: Of course there are left-invariant versions of the mean value theorem (1.33) and the Taylor inequality (1.37): one has merely to replace $f(yx)$ by $f(xy)$, Y^I by X^I, and right Taylor polynomials by left ones. The right-invariant form, however, is the one we shall use most frequently.

If G is stratified, there is a much sharper version of the Taylor inequality which we shall now derive. Again, this result comes in left-invariant and right-invariant forms, but this time it is the left-invariant one which we need most crucially, and therefore the one which we state explicitly.

Suppose then that G is stratified. Let $j = 1,2,\ldots,\nu$ be the indices such that $d_j = 1$, and let V_1 be the linear span of $X_1,\ldots,X_\nu$. Then V_1 generates $\mathfrak{g}$, whence $\exp(V_1)$ generates G. More precisely:

(1.40) LEMMA. <u>If</u> G <u>is stratified, there exist</u> $C > 0$ <u>and</u> $N \in \mathbb{N}$ <u>such that any</u> $x \in G$ <u>can be expressed as</u> $x = x_1 x_2 \cdots x_N$ <u>with</u> $x_j \in \exp(V_1)$ <u>and</u> $|x_j| \le C|x|$ <u>for all</u> j.

<u>Proof</u>: Suppose the lower central series of G terminates at step m, and let $B = \{Y \in V_1 : |\exp Y| \le 1\}$. We define maps $\phi^0, \phi^1_{i_1}, \phi^2_{i_1 i_2}, \ldots, \phi^{m-1}_{i_1 \cdots i_{m-1}}$ $(1 \le i_j \le \nu)$ from B into G by

$$\phi^0(Y) = \exp Y,$$

$$\phi^j_{i_1 \cdots i_j}(Y) = [\cdots[[\exp Y, \exp X_{i_1}], \exp X_{i_2}], \ldots, \exp X_{i_j}]$$

where $[x,y] = xyx^{-1}y^{-1}$. By the Campbell-Hausdorff formula, for any $X, Y \in \mathfrak{g}$ we have

$$[\exp X, \exp Y] = \exp([X,Y] + \text{higher order terms}).$$

Therefore, if we identify the tangent spaces of B and G at the origin with V_1 and $\mathfrak{g}$ respectively, we see that the differential of $\phi^j_{i_1 \cdots i_j}$ at 0 is given by

$$d\phi^0(0)(Y) = Y,$$

$$d\phi^j_{i_1 \cdots i_j}(0)(Y) = [\cdots[[Y, X_{i_1}], X_{i_2}], \ldots, X_{i_j}].$$

Now consider the map

$$\phi((Y^j_{i_1 \cdots i_j})) = \Pi_{j=0}^{m-1} \Pi_{1 \le i_k \le \nu, 1 \le k \le j} \phi^j_{i_1 \cdots i_j}(Y^j_{i_1 \cdots i_j})$$

from the $(\Sigma_0^{m-1} \nu^j)$-fold product of B with itself into G. Since V_1

generates $\mathfrak{g}$, the preceding remarks (together with another application of Campbell-Hausdorff) show that the differential $d\phi(0)$ is surjective onto $\mathfrak{g}$. Consequently, there exists $\delta > 0$ such that the range of ϕ includes all $x \in G$ with $|x| \le \delta$. Since a commutator of $j+1$ elements of G is a product of $3\cdot 2^j-2$ elements, any $x \in G$ with $|x| \le \delta$ can be written as the product of $N = \Sigma_0^{m-1}\nu^j(3\cdot 2^j-2)$ elements of $\exp(V_1)$ whose norms are at most 1. By dilation, then, any $x \in G$ can be written as the product of N elements of $\exp(V_1)$ whose norms are at most $|x|/\delta$. #

If G is stratified, we have $\Delta = \mathbb{N}$, and for $k \in \mathbb{N}$ we define C^k to be the space of continuous functions f on G whose (distribution) derivatives $X^I f$ are continuous functions on G for $d(I) \le k$. In this connection it is worthwhile to note that since V_1 generates $\mathfrak{g}$, the set of left-invariant differential operators which are homogeneous of degree j — that is, the linear span of $\{X^I : d(I) = j\}$ — is precisely the linear span of the operators $X_{i_1} \cdots X_{i_j}$ with $1 \le i_k \le \nu$ for $k = 1,\ldots,j$.

(1.41) STRATIFIED MEAN VALUE THEOREM. <u>Suppose</u> G <u>is stratified. There exist</u> $C > 0$ <u>and</u> $b > 0$ <u>such that for all</u> $f \in C^1$ <u>and all</u> $x,y \in G$,

$$|f(xy)-f(x)| \le C|y|\sup_{|z|\le b|y|,1\le j\le \nu} |X_j f(xz)|.$$

<u>Proof</u>: The proof is identical to the proof of Theorem 1.33 except that one makes the initial estimate only for $y \in \exp(V_1)$ and then uses Lemma 1.40 instead of Lemma 1.31. #

(1.42) THEOREM (STRATIFIED TAYLOR INEQUALITY). *Suppose* G *is stratified. For each positive integer* k *there is a constant* C_k *such that for all* $f \in C^k$ *and all* $x, y \in G$,

$$|f(xy)-P_x(y)| \leq C_k |y|^k \eta(x, b^k|y|),$$

where P_x *is the left Taylor polynomial of* f *at* x *of homogeneous degree* k, b *is as in Theorem 1.41, and for* $r > 0$,

$$\eta(x,r) = \sup_{|z|\leq r, d(I)=k} |X^I f(xz)-X^I f(x)|.$$

Proof: Let $g(x) = f(xy)-P_x(y)$, so that $X^I g(0) = 0$ for $d(I) \leq k$. We shall show by induction on m, $0 \leq m \leq k$, that if $d(J) = k-m$ then

(1.43) $$|X^J g(y)| \leq C_j |y|^m \eta(x, b^m|y|).$$

If $|J| = k$ then $X^J P_x$ is a constant function, hence $X^J P_x(y) = X^J P_x(x) = X^J f(x)$ and so $X^J g(y) = X^J f(xy) - X^J f(x)$. Thus for $m = 0$, (1.43) is just the definition of η. Suppose (1.43) is true for $|J| = k-m+1$, and suppose $|J| = k-m$. Then by Theorem 1.41, since $X^J g(0) = 0$,

$$|X^J g(y)| \leq C|y| \sup_{|z|\leq b|y|, 1\leq j\leq \nu} |X_j X^J g(xz)|.$$

But $X_j X^J$ is a linear combination of X^I's with $D(I) = d(J)+1 = k-m+1$, so by inductive hypotheses,

$$|X^J g(y)| \leq C'|y| \sup_{|z|\leq b|y|} C_{m-1}|z|^{m-1}\eta(x, b^{m-1}|z|)$$

$$\leq C'C_{m-1} b^{m-1} |y|^m \eta(x, b^m|y|).$$

Thus (1.43) is valid for $|J| = k-m$. The desired result follows by taking $J = 0$. #

(1.44) COROLLARY. With notation as above, if $f \in C^{k+1}$ then

$$|f(xy)-P_x(y)| \leq C_k'|y|^{k+1} \sup_{|z|\leq b^{k+1}|y|, d(I)=k+1} |X^I f(xz)|.$$

Proof: Use Theorem 1.41 to estimate $\eta(x, b^k|y|)$. #

(1.45) COROLLARY. If $f \in C^{k+1}$ and $X^I f = 0$ for $d(I) = k+1$ then $f \in P_k$.

Proof: By Corollary 1.44, $f = P_x$ for any $x \in G$. #

D. The Schwartz Class

We define the Schwartz class S on G by identifying G with the Euclidean space $\mathfrak{g}$ via the exponential map. Thus, if $\eta_1, \ldots, \eta_n$ are the canonical coordinates on G introduced in Section C,

$$S = \{\phi \in C^\infty(G): \ P(\partial/\partial\eta)^I\phi \text{ is bounded on } G \text{ for every polynomial } P \text{ and every multiindex } I\}.$$

In view of Proposition 1.25 and the remarks following it, we can replace $(\partial/\partial\eta)^I$ by X^I or Y^I in this definition without changing anything. S is a Fréchet space whose topology is defined by any of a number of families of norms. For our purposes it will be convenient to use the following family: if $N \in \mathbb{N}$, we define

$$\|\phi\|_{(N)} = \sup_{|I|\leq N, x\in G} (1+|x|)^{(N+1)(Q+1)} |Y^I\phi(x)|.$$

Then $\phi_j \to \phi$ in S if and only $\|\phi_j - \phi\|_{(N)} \to 0$ for all N.

If $\phi \in S$ and $y \in G$, let us define

$$\phi^y(x) = \phi(xy), \qquad {}^y\phi(x) = \phi(yx), \qquad \tilde{\phi}(x) = \phi(x^{-1}).$$

(This notation will be used consistently in this section, but not afterwards.)

(1.46) PROPOSITION. <u>For each</u> $N \in \mathbb{N}$ <u>there exists</u> $C_n > 0$ <u>such that for all</u> $\phi \in S$ <u>and</u> $y \in G$,

$$\|\phi^y\|_{(N)} \leq C_N(1+|y|)^{(N+1)(Q+1)}\|\phi\|_{(N)},$$

$$\|{}^y\phi\|_{(N)} \leq C_N(1+|y|)^{(N+1)(Q+1)}\|\phi\|_{(4N)},$$

$$\|\tilde{\phi}\|_{(N)} \leq C_N\|\phi\|_{(2N)}.$$

<u>Moreover</u>, $\|\phi^y - \phi\|_{(N)} \to 0$ <u>and</u> $\|{}^y\phi - \phi\|_{(N)} \to 0$ <u>as</u> $y \to 0$.

<u>Proof</u>: First, by Lemma 1.10,

$$\|\phi^y\|_{(N)} \leq \sup_{|I| \leq N, x \in G} [\gamma(1+|xy|)(1+|y|)]^{(N+1)(Q+1)} |Y^J\phi(xy)|$$

$$\leq [\gamma(1+|y|)]^{(N+1)(Q+1)} \|\phi\|_{(N)}.$$

Next we observe that $(Y^I\tilde{\phi})(x) = (-1)^{|I|}(X^I\phi)(x^{-1})$, and by Proposition 1.29 we have

$$X^I = \sum_{|J| \leq |I|, d(J) \geq d(I)} P_{IJ}Y^J, \qquad P_{IJ} \in P_{d(J)-d(I)}.$$

Now $|J| \leq |I| \leq N$ implies $d(J)-d(I) \leq d(J) \leq \bar{d}|J| \leq QN$, hence $(N+1)(Q+1)+d(J)-d(I) \leq (2N+1)(Q+1)$, and thus

$$\|\tilde{\phi}\|_{(N)} \leq C_N \sup_{|J|\leq N, x\in G} (1+|x|)^{(2N+1)(Q+1)} |Y^J\phi(x^{-1})| \leq C_N\|\phi\|_{(2N)}.$$

The estimate for ${}^y\phi$ then follows immediately since ${}^y\phi = (\tilde{\phi}^{y^{-1}})^{\sim}$. Finally, by the mean value theorem (1.33) and Lemma 1.10,

$$\|{}^y\phi-\phi\|_{(N)} \leq \sup_{|I|\leq N, x\in G, |x|\leq\beta|y|} ((1+|x|)^{(N+1)(Q+1)} \textstyle\sum_1^n |y|^{d_j} |Y_j Y^I\phi(zx)|$$

$$\leq C'(1+|y|)^{(N+1)(Q+1)} \|\phi\|_{(N+1)} \textstyle\sum_1^n |y|^{d_j}$$

$$\longrightarrow 0 \quad \text{as} \quad y \to 0.$$

That $\|\phi^y-\phi\|_{(N)} \to 0$ follows in the same way by using the left-invariant version of the mean value theorem. #

(1.47) PROPOSITION. <u>Convolution is continuous from</u> $\mathcal{S}\times\mathcal{S}$ <u>to</u> $\mathcal{S}$. <u>More precisely, for every</u> $N\in\mathbb{N}$ <u>there exists</u> $C_N > 0$ <u>such that for all</u> $\phi,\psi\in\mathcal{S}$,

$$\|\phi*\psi\|_{(N)} \leq C_N\|\phi\|_{(N)}\|\psi\|_{(N+1)}.$$

<u>Proof</u>: By Lemma 1.10,

$$\|\phi*\psi\|_{(N)} = \sup_{|I|\leq N, x\in G} (1+|x|)^{(N+1)(Q+1)} |(Y^I\phi)*\psi(x)|$$

$$\leq \sup_{|I|\leq N, x\in G} C\int (1+|xy^{-1}|)^{(N+1)(Q+1)} |Y^I\phi(xy^{-1})|(1+|y|)^{(N+1)(Q+1)}|\psi(y)|dy$$

$$\leq C\|\phi\|_{(N)} \int (1+|y|)^{(N+1)(Q+1)}|\psi(y)|dy$$

$$\leq C\|\phi\|_{(N)}\|\psi\|_{(N+1)} \int (1+|y|)^{-Q-1}\,dy. \quad \#$$

The dual space S' of S is the space of <u>tempered distributions</u> on G. If $f \in S'$ and $\phi \in S$ we shall denote the evaluation of f on ϕ by $\langle f,\phi \rangle$ when we wish to be precise; however, we shall usually pretend that distributions are functions and write

$$\langle f,\phi \rangle = \int f(x)\phi(x)dx.$$

Convergence in S' will always mean weak convergence: thus $f_j \to f$ in S' if and only if $\langle f_j,\phi \rangle \to \langle f,\phi \rangle$ for all $\phi \in S$.

If $f \in S'$ and $\phi \in S$ we define the convolution $f*\phi$ by

$$f*\phi(x) = \int f(y)\phi(y^{-1}x)dy = \langle f,(\phi^x)^{\sim} \rangle.$$

It follows easily from Proposition 1.46 that $f*\phi$ is continuous; in fact, it is C^∞ since $X^I(f*\phi) = f*X^I\phi$ for all I.

(1.48) PROPOSITION. *If* $f \in S'$ *there exist* $N \in \mathbb{N}$ *and* $C > 0$ *such that for all* $\phi \in S$ *and* $x \in G$,

$$|f*\phi(x)| \leq C\|\phi\|_{(N)} (1+|x|)^N.$$

In particular, $f*\phi \in S'$, *and for any* $\psi \in S$ *we have*

$$\langle f*\phi,\psi \rangle = \langle f,\psi*\tilde{\phi} \rangle.$$

Proof: The continuity of f as a linear functional on S is equivalent to the existence of $M \in \mathbb{N}$, $C > 0$ such that

$$|\langle f,\phi \rangle| \leq C\|\phi\|_{(M)} \quad \text{for } \phi \in S.$$

Thus by Proposition 1.46, if $N = (2M+1)(Q+1)$ we have

$$|f*\phi(x)| = |\langle f,(\phi^x)^{\sim}\rangle| \le C(1+|x|)^N \|\phi\|_{(2M)} \le C\|\phi\|_{(N)} (1+|x|)^N.$$

The verification of the second statement is a simple exercise which we leave to the reader. #

(1.49) PROPOSITION. <u>Suppose</u> $\phi \in S$ <u>and</u> $\int \phi(x)dx = a$. <u>Then for any</u> $\psi \in S$ <u>and</u> $f \in S'$, $\psi*\phi_t \to a\psi$ <u>in</u> S $f*\phi_t \to af$ <u>in</u> S' <u>as</u> $t \to 0$.

<u>Proof</u>: For the first assertion we merely repeat the proof of Proposition 1.20, using the norms $\|\ \|_{(N)}$ instead of L^p norms; the necessary estimates are provided by Proposition 1.46. The second assertion follows from the first: since $\int \tilde{\phi} = \int \phi$ and $(\tilde{\phi})_t = (\phi_t)^{\sim}$, for any $\psi \in S$ we have

$$\langle f*\phi_t,\psi\rangle = \langle f,\psi*\tilde{\phi}_t\rangle \to a\langle f,\psi\rangle \quad \text{as } t \to 0. \quad \#$$

The remainder of this section is devoted to some technical results which we shall need later. The first one is a global version of the Taylor inequality (1.37) for Schwartz class functions.

(1.50) THEOREM. <u>Suppose that</u> $a \in \Delta$, $b = \min\{b' \in \Delta : b' > a\}$, <u>and</u> $N = [a]+1$. <u>If</u> $\phi \in S$ <u>and</u> $x \in G$, <u>let</u> P_x <u>be the right Taylor polynomial of</u> ϕ <u>at</u> x <u>of homogeneous degree</u> a, <u>and let</u> $R_x(y) = \phi(yx)-P_x(y)$. <u>There is a constant</u> C, <u>independent of</u> ϕ, <u>such that</u>

$$|Y^I R_x(y)| \le C\|\phi\|_{(N)} |y|^{b-d(I)} |x|^{-Q-b}$$

<u>whenever</u> $d(I) \le a$ <u>and</u> $|x| \ge 2\gamma\beta^N|y|$.

Proof: If $d(I) \leq a$ then $Y^I P_x$ is the right Taylor polynomial of $Y^I \phi$ at x of homogeneous degree $a-d(I)$, and

$$[a-d(I)] \leq [a-|I|] = [a]-|I| = N-1-|I|.$$

Thus by the Taylor inequality (1.37),

$$|Y^I R_x(y)| \leq C \sum_{|J|+|I|\leq N, d(J)+d(I)\geq b} |y|^{d(J)} \sup_{|z|\leq \beta^{N-|I|}|y|} |Y^J Y^I \phi(zx)|.$$

Now $|z| \leq \beta^N|y|$ implies $|z| \leq |x|/2\gamma$ and hence $|zx| \geq |x|/2$, and $Y^J Y^I$ is a linear combination of Y^K's with $d(K) = d(J)+d(I)$ and $|K| \leq |I|+|J| \leq N$. Therefore,

$$\sup_{|z|\leq \beta^{N-|I|}|y|} |Y^J Y^I \phi(zx)| \leq \sup_{|z|\leq \beta^N |y|} \|\phi\|_{(N)} (1+|zx|)^{-QN-Q}$$

$$\leq 2^{QN+Q}\|\phi\|_{(N)} (1+|x|)^{-QN-Q},$$

and hence, since $|I|+|J| \leq N$ implies $d(I)+d(J) \leq QN$,

$$|Y^I R_x(y)| \leq C'\|\phi\|_{(N)}|y|^{b-d(I)} [\sum_{|J|+|I|\leq N, d(J)+d(I)\geq b}$$

$$(|y|/|x|)^{d(J)+d(I)-b} (1+|x|)^{-QN-Q+d(J)+d(I)-b}]$$

$$\leq C''\|\phi\|_{(N)}|y|^{b-d(I)}|x|^{-Q-b}. \quad \#$$

The next sequence of lemmas deals with estimates for functions of the form $\psi*\partial_s^j\phi_s$ where $\phi,\psi \in \mathcal{S}$, $0 < s \leq 1$, $\partial_s = \partial/\partial s$, and (as usual), $\phi_s = s^{-Q}\phi \circ \delta_{1/s}$. We consider ϕ as fixed, and we are interested in estimating $\psi*\partial_s^j\phi_s$ in terms of ψ. With a little more work we could display explicitly the dependence of these estimates on ϕ, but we shall have no need to do so.

(1.51) LEMMA. Given $j \in \mathbb{N}$, there exist polynomials $P_{iI}, Q_{iI} \in \mathcal{P}_{j(2\bar{d}-1)}$ $(0 \leq i \leq j(2\bar{d}-1),\ 0 \leq |I| \leq j)$ such that

$$\psi*\partial_s^j\phi_s = \sum_{i=0}^{j(2\bar{d}-1)} \sum_{|I|\leq j} s^i P_{iI}[Y^I\psi*(Q_{iI}\phi)_s].$$

Proof: We have

$$\psi*\partial_s^j\phi_s(x) = \partial_s^j(\psi*\phi_s)(x) = \partial_s^j \int \psi(xy^{-1})s^{-Q}\phi(y/s)dy$$

$$= \partial_s^j \int \psi(x(sy^{-1}))\phi(y)dy.$$

Let $\eta_1,\ldots,\eta_n$ be the canonical coordinates on G defined in Section C. We differentiate under the integral sign, using the chain rule

$$\partial_s[\psi(x(sy^{-1}))] = \sum \frac{\partial\psi}{\partial\eta_k} \frac{\partial\eta_k(x(sy^{-1}))}{\partial s}$$

and then express the $(\partial/\partial\eta)$'s in terms of the vector fields $Y_1,\ldots,Y_n$. The derivative $\partial\eta_k(x(sy^{-1}))/\partial s$ is a polynomial of homogeneous degree $d_k \leq \bar{d}$ in x and y and of degree $d_k-1 \leq \bar{d}-1$ in s, and $\partial/\partial\eta_j = \Sigma P_{k\ell}Y_\ell$ where $P_{k\ell} \in \mathcal{P}_{d_\ell-d_k} \subset \mathcal{P}_{\bar{d}-1}$. $P_{k\ell}$, moreover, is to be evaluated at $x(sy^{-1})$, so if we apply ∂_s j times and regroup terms, we find that

$$\psi*\partial_s^j\phi_s(x) = \int \sum_{i=0}^{j(2\bar{d}-1)} \sum_{|I|\leq j} s^i P_{iI}(x)Q_{iI}(y)Y^I\psi(x(sy^{-1}))\phi(y)dy,$$

where P_{iI}, Q_{iI} are of homogeneous degree $\leq j(2\bar{d}-1)$. But this is precisely what we wished to show. #

(1.52) LEMMA. *For all* $N, j \in \mathbb{N}$ *there exists* $C > 0$ *such that*

$$\sup_{0<s\leq 1} \int (1+|x|)^N |\psi * \partial_s^j \phi_s(x)| dx \leq C \sup_{x\in G, |I|\leq j} (1+|x|)^{N'} |Y^I \psi(x)|$$

where $N' = N+Q+1+j(2\bar{d}-1)$.

Proof: By Lemma 1.51 and Lemma 1.10, for $0 < s \leq 1$ we have

$$\int (1+|x|)^N |\psi * \partial_s^j \phi_s(x)| dx$$

$$\leq \left(\int (1+|x|)^{-Q-1} dx\right) \sup_{x\in G} (1+|x|)^{N+Q+1} |\psi * \partial_s^j \phi_s(x)|$$

$$\leq \sup_{x\in G} C(1+|x|)^{N'} \Sigma_{i,I} |Y^I \psi * (Q_{iI}\phi)_s(x)|$$

$$\leq \sup_{x\in G} C' \Sigma_{i,I} \int (1+|xy^{-1}|)^{N'} |Y^I \psi(xy^{-1})| (1+|y|)^{N'} |(Q_{iI}\phi)_s(y)| dy$$

$$\leq \sup_{x\in G} C' \Sigma_{i,I} (1+|x|)^{N'} |Y^I \psi(x)| \int (1+s|y|)^{N'} |(Q_{iI}\phi)(y)| dy$$

$$\leq C'' \sup_{x\in G} (1+|x|)^{N'} |Y^I \psi(x)|. \quad \#$$

(1.53) LEMMA. *For every* $N, j \in \mathbb{N}$ *and every multiindex* I *there exists* $C > 0$ *such that*

$$\sup_{x\in G, 0<s\leq 1} (1+|x|)^N |Y^I (\psi * \partial_s^j \phi_s)(x)|$$

$$\leq C \sup_{x\in G, |J|\leq j+|I|} (1+|x|)^{N'} |Y^J \psi(x)|$$

where $N' = N+j(2\bar{d}-1)$.

Proof: Since $Y^I(\psi * \partial_s^j \phi_s) = (Y^I\psi)*(\partial_s^j\phi_s)$, we merely repeat the proof of Lemma 1.52, omitting the first step and replacing ψ by $Y^I\psi$. #

(1.54) LEMMA. For all $N, j_1, j_2, \ldots, j_k \in \mathbb{N}$ there exists $C > 0$ such that

$$\sup_{0<s_1,s_2,\ldots,s_k\leq 1} \int (1+|x|)^N |\psi * \partial_{s_1}^{j_1}\phi_{s_1} * \cdots * \partial_{s_k}^{j_k}\phi_{s_k}(x)|dx$$

$$\leq C \sup_{x\in G, |J|\leq \Sigma j_i} (1+|x|)^{N'} |Y^J\psi(x)|,$$

where $N' = N+Q+1+(\Sigma_1^k j_i)(2\bar{d}-1)$.

Proof: The case $k = 1$ is Lemma 1.52. If $k = 2$, for $0 < s_1 \leq 1$ we set $\psi' = \psi * \partial_{s_1}^{j_1}\phi_{s_1}$. By Lemmas 1.52 and 1.53,

$$\sup_{0<s_2\leq 1} \int (1+|x|)^N |\psi * \partial_{s_1}^{j_1}\phi_{s_1} * \partial_{s_2}^{j_2}\phi_{s_2}(x)|dx$$

$$\leq C \sup_{x\in G, |I|\leq j_2} (1+|x|)^{N+Q+1+j_2(2\bar{d}-1)} |Y^I\psi'(x)|$$

$$\leq C' \sup_{x\in G, |I|\leq j_2, |J|\leq j_1} (1+|x|)^{N+Q+1+(j_1+j_2)(2\bar{d}-1)} |Y^J Y^I\psi(x)|$$

$$\leq C'' \sup_{x\in G, |K|\leq j_1+j_2} (1+|x|)^{N+Q+1+(j_1+j_2)(2\bar{d}-1)} |Y^K\psi(x)|.$$

Since this holds for all $s_1 \in (0,1]$, the assertion is valid for $k = 2$. The proof is now completed by an obvious induction on k.

(1.55) PROPOSITION. <u>Given</u> $\phi \in S$ <u>and</u> $N, j_1, \ldots, j_k \in \mathbb{N}$ <u>with</u> $\Sigma_1^k j_i = N+1$, there exists $C > 0$ such that for all $\psi \in S$,

$$(1.56) \qquad \sup_{0<s\leq 1} \int (1+|x|)^N |\partial_s^{j_1}\phi_s * \cdots * \partial_s^{j_k}\phi_s * \psi(x)|\,dx \leq C\|\psi\|_{(3N+3)}.$$

<u>Proof</u>: We observe that if $\tilde{f}(x) = f(x^{-1})$ then $(f*g)^{\sim} = \tilde{g}*\tilde{f}$ and $(f_s)^{\sim} = (\tilde{f})_s$. Hence if we set $x = y^{-1}$ in (1.56), the left hand side becomes

$$\sup_{0<s\leq 1} \int (1+|y|)^N |\tilde{\psi}*\partial_s^{j_k}\tilde{\phi}_s * \cdots * \partial_s^{j_1}\tilde{\phi}_s(y)|\,dy$$

which, by Lemma 1.54 with $s_1 = \cdots = s_k = s$, is dominated by

$$\sup_{x\in G, |J|\leq N+1} (1+|x|)^{N+Q+1+(N+1)(2\bar{d}-1)} |Y^J\tilde{\psi}(x)|.$$

But by Proposition 1.29,

$$Y^J\tilde{\psi}(x) = (-1)^{|J|}X^J\psi(x^{-1}) = \sum_{|K|\leq|J|, d(K)\geq d(J)} P_{JK}(x^{-1})Y^J\psi(x^{-1})$$

where $P_{JK} \in \mathcal{P}_{d(K)-d(J)}$. Since $|K| \leq |J|$ implies $d(K)-d(J) \leq (\bar{d}-1)|J|$, the left hand side of (1.56) is dominated by

$$(1.57) \qquad \sup_{x\in G, |J|\leq N+1} (1+|x|)^{N+Q+1+(N+1)(3\bar{d}-2)} |Y^J\psi(x)|.$$

Finally, since $\bar{d} \leq Q$, we have

$$N+Q+1+(N+1)(3\bar{d}-2) \leq Q+(N+1)(3Q) \leq (3N+4)(Q+1),$$

so (1.57) is dominated by $\|\psi\|_{(3N+3)}$. #

By essentially the same argument we also obtain the following result (which of course can be made into a quantitative estimate):

(1.58) PROPOSITION. If $\phi,\psi \in S$ and $j_1,\dots,j_k \in \mathbb{N}$ then

$$\sup_{x\in G,\,0<s\leq 1} |\partial_s^{j_1}\phi_s * \cdots * \partial_s^{j_k}\phi_s * \psi(x)| < \infty .$$

E. Integral Representations of the δ Function

The next group of results concerns the possibility of expressing the Dirac δ-function or point mass at the origin in the form

$$(1.59) \qquad \delta = \int_0^\infty \phi_t \, dt/t$$

where ϕ is a sum of convolutions of Schwartz class functions with many vanishing moments. Equation (1.59) is always to be interpreted to mean

$$(1.59') \qquad \delta = \lim_{\varepsilon\to 0,\, A\to\infty} \int_\varepsilon^A \phi_t \, dt/t \qquad \text{(convergence in } S'\text{)}.$$

(1.60) LEMMA. Suppose $M,N \in \mathbb{N}$, $M \geq [\bar{d}]N$, and $\phi \in S$ satisfies $\int \phi P = 0$ for all $P \in P_M^{iso}$. Then there exist $\phi_J \in S$ $(N \leq d(J) \leq \bar{d}N)$ such that $\phi = \sum_{N\leq d(J)\leq \bar{d}N} X^J \phi_J$ and $\int \phi_J P = 0$ for all $P \in P_{M-[\bar{d}]N}^{iso}$. Similarly, there exist $\psi_J \in S$ such that $\phi = \sum_{N\leq d(J)\leq \bar{d}N} Y^J \psi_J$ and $\int \psi_J P = 0$ for all $P \in P_{M-[\bar{d}]N}^{iso}$.

Proof: First consider the special case $N = 1$, $G = \mathbb{R}^n$, $X_j = \partial/\partial x_j$, $d_j = 1$. Then, by taking Fourier transforms, the problem can be rephrased as follows: given $\phi \in \mathcal{S}$ such that $\hat{\phi}$ vanishes to order M at the origin, find $\phi_1,\ldots,\phi_n \in \mathcal{S}$ such that $\hat{\phi}(\xi) = \Sigma_1^n \xi_j \hat{\phi}_j(\xi)$ and $\hat{\phi}_j$ vanishes to order $M-1$ at the origin. But this is easy: for $|\xi| \leq 2$ we may write

$$\hat{\phi}(\xi) = \int_0^1 (d/dt)\hat{\phi}(t\xi)dt = \Sigma_1^n \int_0^1 \xi_j(\partial\hat{\phi}/\partial\xi_j)(t\xi)dt,$$

while for $|\xi| \geq 1$ we may write

$$\hat{\phi}(\xi) = \Sigma_1^n \xi_j(\xi_j/|\xi|^2)\hat{\phi}(\xi).$$

Hence if we pick $\zeta \in C_0^\infty(B(2,0))$ with $\zeta = 1$ on $B(1,0)$, we can take

$$\hat{\phi}_j(\xi) = \zeta(\xi)\int_0^1 (\partial\hat{\phi}/\partial\xi_j)(t\xi)dt + (1-\zeta(\xi))(\xi_j/|\xi|^2)\hat{\phi}(\xi).$$

Next, we consider the case $N = 1$ on a general homogeneous group G. We reduce this to the case $G = \mathbb{R}^n$ by using canonical coordinates $\eta_1,\ldots,\eta_n$ as in Section C. By the above argument we have $\phi = \Sigma_1^n(\partial\phi_j/\partial\eta_j)$ where $\int \phi_j P = 0$ for all $P \in \mathcal{P}_{M-1}^{iso}$. But there exist $P_{jk} \in \mathcal{P}_{d_k - d_j}$ (so that $X_k P_{jk} = 0$) such that for all $\psi \in \mathcal{S}$,

$$\partial\psi/\partial\eta_j = \Sigma_k P_{jk} X_k \psi = \Sigma_k X_k(P_{jk}\psi).$$

Hence

$$\phi = \Sigma(\partial\phi_j/\partial\eta_j) = \Sigma X_k \phi_k' \quad \text{where} \quad \phi_k' = \Sigma P_{jk}\phi_j.$$

Moreover, since $\mathcal{P}_{d_k-d_j} \subset \mathcal{P}_{\bar{d}-1} \subset \mathcal{P}^{iso}_{[\bar{d}]-1}$ by Lemma 1.36, $\int \phi_k' P = 0$ for all $P \in \mathcal{P}^{iso}_{M-1-([\bar{d}]-1)} = \mathcal{P}^{iso}_{M-[\bar{d}]}$, so we have proved the case $N = 1$.

The general case is now established by induction on N. If $\phi \in S$ and $\int \phi P = 0$ for all $P \in \mathcal{P}^{iso}_M$ then $\phi = \Sigma X_j \phi_j$ where $\int \phi_j P = 0$ for all $P \in \mathcal{P}^{iso}_{M-[\bar{d}]}$. By inductive hypothesis, $\phi_j = \sum_{N-1 \le d(I) \le \bar{d}(N-1)} X^I \phi_{Ij}$ where $\int \phi_{Ij} P = 0$ for all $P \in \mathcal{P}^{iso}_{M-[\bar{d}]N}$, from which the desired result follows immediately. #

(1.61) THEOREM. For any $N \in \mathbb{N}$ there exist $\phi^1, \ldots, \phi^M, \psi^1, \ldots, \psi^M \in S$ (where M depends on N) such that:

(a) $\int \phi^j P = \int \psi^j P = 0$ for all $P \in \mathcal{P}_N$, $1 \le j \le M$,

(b) $\sum_1^M \int_0^\infty \phi_t^j * \psi_t^j \, dt/t = \delta$.

Proof: First pick $\phi \in S$ such that $\int \phi = 1$ and $\int \phi P = 0$ for every polynomial P without constant term. (For example, let ϕ be the inverse [Euclidean] Fourier transform of a Schwartz class function which is identically one near the origin.) Set $\phi' = d\phi_t/dt|_{t=1}$. Since ϕ_t satisfies the same conditions as ϕ, it follows easily that $\int \phi' P = 0$ for every polynomial P. Moreover, $\int \phi * \phi = 1$, $(\phi * \phi)_t \to 0$ uniformly as $t \to \infty$, and $d\phi_t/dt = \phi_t'/t$.

Hence by Proposition 1.49,

$$\delta = \lim_{\varepsilon\to 0, A\to\infty} [(\phi*\phi)_\varepsilon - (\phi*\phi)_A] = -\lim\int_\varepsilon^A (d/dt)(\phi_t*\phi_t)dt$$

$$= -\lim\int_\varepsilon^A (\phi_t*\phi_t' + \phi_t'*\phi_t)dt/t.$$

By Lemma 1.60 (with N replaced by $N+1$ and $M \geq N + [\bar{d}](N+1)$) we can write

$$\phi' = \sum_{N+1\leq d(I)\leq\bar{d}(N+1)} X^I\psi^I = \sum_{N+1\leq d(I)\leq\bar{d}(N+1)} Y^I\theta^I$$

where $\int\psi^I P = \int\theta^I P = 0$ for all $P \in \mathcal{P}_N \subset \mathcal{P}_N^{iso}$. Hence

$$\phi*\phi' = \phi*(\Sigma Y^I\theta^I) = \Sigma(X^I\phi)*\theta^I,$$

$$\phi'*\phi = (\Sigma X^I\psi^I)*\phi = \Sigma\psi^I*(Y^I\phi).$$

But now we are done, since if $P \in \mathcal{P}_N$ and $d(I) \geq N+1$,

$$\int(X^I\phi)P = \pm\int\phi(X^I P) = 0 = \pm\int\phi(Y^I P) = \int(Y^I\phi)P. \quad \#$$

Our next result shows that it is possible to assume in Theorem 1.61 that the ψ^j's have compact support.

(1.62) THEOREM. *Let* N, ϕ^j, ψ^j *be as in Theorem* 1.61, *and let* $N' = [[N/\bar{d}]/[\bar{d}]]$. *There exist* $\Phi^1,\ldots,\Phi^{M'}$, $\Psi^1,\ldots,\Psi^{M'} \in \mathcal{S}$ (*where* M' *depends on* N) *such that*

(a) $\int\Phi^k P = \int\Psi^k P = 0$ *for all* $P \in \mathcal{P}_{N'-1}$,

(b) _each_ Φ^k _is of the form_ $\phi^j * \alpha$ _with_ $\alpha \in S$ _and_ $1 \leq j \leq M$,

(c) _each_ Ψ^k _is supported in_ $B(1,0)$,

(d) $\Sigma_1^m \phi^j * \psi^j = \Sigma_1^{M'} \Phi^k * \Psi^k$.

Proof: Since $P^{iso}_{[N/\bar{d}]} \subset P_N$, by Lemma 1.60 (with M,N replaced by $[N/\bar{d}]$, N' respectively) we have

$$\psi^j = \sum_{N' \leq d(I) \leq \bar{d}N'} X^I \psi^{jI} \qquad (\psi^{jI} \in S).$$

Moreover, by Theorem 7.2 of Dixmier and Malliavin [1] we can write

$$\psi^{jI} = \sum_{k=1}^K \alpha^{iJk} * \beta^{iJk} \quad \text{where} \quad \alpha^{iJK} \in S, \quad \beta^{iJk} \in C_0^\infty(B(1,0)).$$

But then

$$\Sigma_j \phi^j * \psi^j = \Sigma_{jIk} \Phi^{jIk} * \Psi^{jIk}$$

where

$$\Phi^{jIk} = \phi^j * \alpha^{jIk}, \qquad \Psi^{jIk} = X^I \beta^{jIk}.$$

We have $\int \Phi^{jIk} P = 0$ for all $P \in P_N$ since the same is true of ϕ^j, while $\int \Psi^{jIk} P = \pm \int \beta^{jIk}(X^I P) = 0$ for $P \in P_{N'-1}$ since $d(I) \geq N'$. #

If $\int_0^\infty \phi_t dt/t = \delta$ and $\psi \in S$ then

(1.63) $$\int_\varepsilon^A \psi * \phi_t dt/t \to \psi \text{ in } S' \text{ as } \varepsilon \to 0, A \to \infty.$$

This relation need not hold if ψ is merely a distribution: for example, if ψ is a nonzero constant function and $\int \phi = 0$ (which is the case in the examples we have constructed, and indeed is necessary for the convergence of $\int_0^\infty \phi_t dt/t$ at $t = 0$), the expression on the left of (1.63) vanishes identically. However, we shall now show that (1.63) still holds for certain kinds of distributions.

In the following arguments we regard functions on G as functions on $\mathfrak{g}$ via the exponential map. We then have the Fourier transform

$$\hat{\phi}(\xi) = \int e^{-i\langle \xi, x\rangle} \phi(x)dx,$$

which is an isomorphism from $S(\mathfrak{g})$ and $S'(\mathfrak{g})$ to $S(\mathfrak{g}^*)$ and $S'(\mathfrak{g}^*)$ respectively ($\mathfrak{g}^*$ = dual space of $\mathfrak{g}$). Moreover, the dilations on $\mathfrak{g}$ induce dilations on $\mathfrak{g}^*$ which are given by $t(\xi_1, \ldots, \xi_n) = (t^{d_1}\xi_1, \ldots, t^{d_n}\xi_n)$, where $\xi_1, \ldots, \xi_n$ are the coordinates on $\mathfrak{g}^*$ induced by the basis $X_1, \ldots, X_n$ of $\mathfrak{g}$. It is then easily verified that for $\phi \in S$ we have $(\phi_t)\hat{}(\xi) = \hat{\phi}(t\xi)$.

If $f \in S'$, we say that f <u>vanishes weakly at infinity</u> if, for any $\phi \in S$, $f*\phi_t \to 0$ in S' as $t \to \infty$. For example, if $f \in L^p$ where $1 \le p < \infty$, then f vanishes weakly at infinity, since if q is the conjugate exponent to p,

$$\|f*\phi_t\|_\infty \le \|f\|_p \|\phi_t\|_q = \|f\|_p \|\phi\|_q t^{-Q/p}.$$

(1.64) THEOREM. *Suppose* $\phi \in S$, $\int \phi = 0$, *and* $\int_0^\infty \phi_t dt/t = \delta$. *Then for any* $f \in S'$ *which vanishes weakly at infinity,*

$$\int_\varepsilon^A f*\phi_t dt/t \to f \quad \text{in } S' \text{ as } \varepsilon \to 0,\ A \to \infty.$$

<u>Proof</u>: Let $\alpha = \int_0^1 \phi_t dt/t$, $\beta = \int_1^\infty \phi_t dt/t$. Since $\int \phi = \hat{\phi}(0) = 0$, we can write $\hat{\phi}(\xi) = \Sigma_1^n \xi_j \hat{\phi}_j(\xi)$ with $\phi_j \in S$ as in the proof of Lemma 1.60. Thus

$$\hat{\alpha}(\xi) = \Sigma_1^n \int_0^1 t^{d_j - 1} \xi_j \hat{\phi}_j(t\xi) dt,$$

$$(\partial/\partial\xi)^I \hat{\alpha}(\xi) = \int_0^1 t^{d(I)-1} [(\partial/\partial\xi)^I \hat{\phi}](t\xi) dt \qquad (I \neq 0).$$

The integrals on the right converge uniformly in ξ, so $\hat{\alpha}$ is C^∞. Also $\hat{\alpha}(0) = 0$. On the other hand, since $|(\partial/\partial\xi)^I \hat{\phi}(\xi)| \leq C_N(1+|\xi|)^{-N}$ for any N, it is easily seen that the integral $\int_1^\infty \hat{\phi}(t\xi)dt/t$ and all of its derivatives converge uniformly on the set where $|\xi| \geq c$, for any $c > 0$, and are rapidly decreasing as $\xi \to \infty$. In other words, $\hat{\beta}$ agrees with a Schwartz class function except perhaps near the origin. But $\hat{\beta} = \hat{\delta} - \hat{\alpha} = 1 - \hat{\alpha}$, so $\hat{\beta}$ is also smooth near the origin. Therefore $\hat{\beta}$, and hence also β, is in S, and $\int \beta = 1 - \int \alpha = 1 - \hat{\alpha}(0) = 1$. Now observe that for $s > 0$,

$$\beta_s = \int_1^\infty \phi_{st} dt/t = \int_s^\infty \phi_t dt/t,$$

so that

$$\int_\varepsilon^A \phi_t dt/t = \beta_\varepsilon - \beta_A.$$

If $f \in S'$, then, $\int_\varepsilon^A f*\phi_t dt/t = f*\beta_\varepsilon - f*\beta_A$. But $f*\beta_\varepsilon \to f$ as $\varepsilon \to 0$ by Proposition 1.49, and if f vanishes weakly at infinity, $f*\beta_A \to 0$ as $A \to \infty$. #

Our final result shows that if $\phi \in S$ and $\int \phi = 0$ then $\int_0^\infty \phi_t dt/t$ always converges in S', although usually not to the δ-function.

(1.65) THEOREM. If $\phi \in S$ and $\int \phi = 0$ then $\int_\varepsilon^A \phi_t dt/t$ converges in S' as $\varepsilon \to 0$, $A \to \infty$ to a distribution which is C^∞ away from the origin and homogeneous of degree $-Q$.

Proof: As in the proof of Theorem 1.64 we write $\hat{\phi}(\xi) = \Sigma_1^n \xi_j \hat{\phi}_j(\xi)$ with $\phi_j \in S$. Then, if $N > Q+\bar{d}$ and $|\xi|$ denotes a homogeneous norm on g^*,

$$\begin{aligned}\int_0^\infty |\hat{\phi}(t\xi)|dt/t &= \int_0^\infty |\Sigma_1^n t^{d_j-1} \xi_j \hat{\phi}(t\xi)|dt \\ &\leq C\Sigma_1^n \int_0^\infty t^{d_j-1} |\xi|^{d_j}(1+|t\xi|)^{-N}dt \\ &\leq C\Sigma_1^n \left[\int_0^{1/|\xi|} t^{d_j-1} |\xi|^{d_j} dt + \int_{1/|\xi|}^\infty t^{d_j-N-1} |\xi|^{d_j-N} dt\right] \\ &= C'.\end{aligned}$$

This shows that $\int_0^\infty \hat{\phi}(t\xi)dt/t$ converges pointwise and boundedly, hence in S'. Therefore $\int_0^\infty \phi_t dt/t$ converges in S'. Moreover, for any $\psi \in S$ and $r > 0$,

$$\int_G \int_\varepsilon^A \phi_t(x)\psi(rx)(dt/t)dx = \int_G \int_\varepsilon^A \phi_{rt}(x)\psi(x)(dt/t)dx = \int_G \int_{\varepsilon/r}^{A/r} \phi_t(x)\psi(x)(dt/t)dx$$

Letting $\varepsilon \to 0$, $A \to \infty$, we see from the definition of homogeneity for distributions that $\int_0^\infty \phi_t dt/t$ is homogeneous of degree $-Q$. Finally, if

$K \subset G$ is a compact set which does not contain the origin and I is a multiindex, $X^I\phi(x/t)$ vanishes to infinite order as $t \to 0$ and remains bounded as $t \to \infty$, uniformly for $x \in K$, so the integrals

$$X^I \int_0^\infty \phi_t(x)dt/t = \int_0^\infty t^{-Q-d(I)-1} X^I\phi(x/t)dt$$

converge uniformly on K. Thus $\int_0^\infty \phi_t dt/t$ is C^∞ away from the origin. #

F. Covering Lemmas

In this section we present two useful covering lemmas, which are variants of classical results on $\mathbb{R}^n$ due to Wiener and Whitney.

(1.66) WIENER LEMMA. *Suppose* $E \subset G$ *and* $r : E \to (0,\infty)$ *is an arbitrary positive function. Assume that either* (a) E *is bounded, or* (b) E *is open,* $|E| < \infty$, *and* $B(r(x),x) \subset E$ *for all* $x \in E$. *Then there exists a (finite or infinite) sequence* $\{x_j\}$ *in* E *such that the balls* $B(r(x_j),x_j)$ *are disjoint, and* $E \subset \bigcup_j B(4\gamma r(x_j),x_j)$.

Proof: We may assume that $\sup_{x\in E} r(x) < \infty$. In case (b) this is automatic, whereas in case (a), if $\sup_{x\in E} r(x) = \infty$ there exists $x \in E$ such that $E \subset B(r(x),x)$, so there is nothing to prove. Pick $x_1 \in E$ such that $r(x_1) \geq 1/2 \sup_{x\in E} r(x)$. If $E \subset B(4\gamma r(x_1),x_1)$ we are done. Otherwise, we continue inductively: having picked $x_1,\ldots,x_j$, we set $E_j = E\setminus\bigcup_1^j B(4\gamma r(x_i),x_i)$. If $E_j = \emptyset$ we stop. If not, we pick $x_{j+1} \in E_j$ such that $r(x_{j+1}) \geq 1/2 \sup_{x\in E_j} r(x)$. Observe that if $i < j$ then $r(x_j) \leq 2r(x_i)$ (otherwise, we made the wrong choice of x_i). Hence if

$B(r(x_j),x_j)$ intersects $B(r(x_i),x_i)$, we have

$$4\gamma r(x_i) \leq |x_i^{-1}x_j| \leq \gamma(r(x_i)+r(x_j)) \leq 3\gamma r(x_i)$$

which is a contradiction. Hence the balls $B(r(x_j),x_j)$ are disjoint.

We claim that the balls $B(4\gamma r(x_j),x_j)$ cover E. If the sequence $\{x_j\}$ is finite this follows from our construction. If not, since the balls $B(r(x_j),x_j)$ are disjoint and contained in a set of finite measure, we must have $r(x_j) \to 0$ as $j \to \infty$. Hence if there existed $x \in E \backslash \bigcup_1^\infty B(4\gamma r(x_j),x_j)$ we would have $r(x) > 2r(x_k)$ for k sufficiently large, contradicting the choice of x_k. #

(1.67) WHITNEY LEMMA. <u>Suppose</u> E <u>is an open set of finite measure in</u> G, <u>and</u> $C \geq 1$. <u>There exist</u> $x_1,x_2,\ldots$ <u>in</u> E <u>and positive numbers</u> $r_1,r_2,\ldots$ <u>such that</u>:

(a) $E = \bigcup_j B(r_j,x_j)$,

(b) <u>the balls</u> $B(r_j/4\gamma,x_j)$ <u>are disjoint</u>,

(c) $B(Cr_j,x_j) \cap E^c = \emptyset$, <u>but</u> $B(3\gamma Cr_j,x_j) \cap E^c \neq \emptyset$,

(d) <u>no point of</u> E <u>belongs to more than</u> M <u>of the balls</u> $B(Cr_j,x_j)$, <u>where</u> M <u>is the greatest integer in</u> $[8C\gamma^3(1+2\gamma)]^Q$.

<u>Proof</u>: If $x \in E$, let $\rho(x,E^c) = \inf\{|x^{-1}y| : y \in E^c\}$, and let $s(x) = (8C\gamma^2)^{-1}\rho(x,E^c)$. Since E is open, $s(x) > 0$ for all $x \in E$. Thus by the Wiener lemma (1.66), there is a sequence $\{x_j\}$ in E such that the balls $B(s(x_j),x_j)$ are disjoint, while the balls $B(4\gamma s(x_j),x_j)$ cover E.

Let $r_j = 4\gamma s(x_j) = (2C\gamma)^{-1}\rho(x,E^c)$. Then clearly (a), (b), and (c) are satisfied. To prove (d), suppose $x \in E$, and set $R = \rho(x,E^c)$. If $x \in B(Cr_j,x_j)$ then

$$2C\gamma r_j = \rho(x_j,E^c) \leq \gamma(|x^{-1}x_j| + \rho(x,E^c)) < C\gamma r_j + \gamma R,$$

so that $Cr_j < R$. Hence for any $y \in B(Cr_j,x_j)$,

$$|x^{-1}y| \leq \gamma(|x^{-1}x_j| + |x_j^{-1}y|) \leq 2\gamma Cr_j < 2\gamma R.$$

In other words, $B(Cr_j,x_j) \subset B(2\gamma R,x)$. On the other hand,

$$R = \rho(x,E^c) \leq \gamma(|x^{-1}x_j| + \rho(x_j,E^c)) < \gamma(Cr_j+2C\gamma r_j) = C\gamma r_j(1+2\gamma),$$

so $r_j > R/C\gamma(1+2\gamma)$. In short, if x lies in M of the balls $B(Cr_j,x_j)$ then there are M disjoint balls of radius at least $R/4C\gamma^2(1+2\gamma)$ contained in $B(2\gamma R,x)$ (namely, the corresponding balls $B(r_j/4\gamma,x_j)$). But then

$$(2\gamma R)^Q = |B(2\gamma R,x)| \geq M[R/4C\gamma^2(1+2\gamma)]^Q,$$

which implies that $M \leq [8C\gamma^3(1+2\gamma)]^Q$. #

G. The Heat Kernel on Stratified Groups

In this section we assume that G is a stratified group. On such groups there is a natural analogue of the Gaussian kernel on $\mathbb{R}^n$, which plays an important role in analysis.

As in Section C, we let $j = 1,\ldots,\nu$ be those indices for which $d_j = 1$, and we define the <u>sub-Laplacian</u> L of G by

$$L = -\Sigma_1^\nu X_j^2.$$

The *heat operator* associated to L is the differential operator $\partial_t + L$ on $G \times \mathbb{R}$, where $\partial_t = \partial/\partial t$ is the coordinate vector field on $\mathbb{R}$. By a celebrated theorem of Hörmander [1], L and $\partial_t + L$ are both hypoelliptic. That is, if u is a distribution on G (resp. $G \times \mathbb{R}$) such that Lu (resp. $(\partial_t + L)u$) is C^∞ on some open set Ω, then u must be C^∞ on Ω.

(1.68) PROPOSITION. *There is a unique* C^∞ *function* h *on* $G \times (0,\infty)$ *with the following properties:*

(i) $(\partial_t + L)h = 0$ *on* $G \times (0,\infty)$.

(ii) $h(x,t) \geq 0$, $h(x,t) = h(x^{-1},t)$, and $\int h(y,t)dy = 1$ *for all* $x \in G$, $t > 0$.

(iii) $h(\cdot,s)*h(\cdot,t) = h(\cdot,s+t)$ *for all* $s,t > 0$.

(iv) $h(rx,r^2t) = r^{-Q}h(x,t)$ *for all* $x \in G$, $t > 0$, $r > 0$.

Proof: By a theorem of G. Hunt [1], the operator L determines a unique family $\{\mu_t\}_{t>0}$ of probability measures on G such that $\mu_s * \mu_t = \mu_{s+t}$ for all $s,t > 0$ and such that for every $u \in C_0^\infty(G)$, $\partial_t(u*\mu_t) = -(Lu)*\mu_t$. Moreover, the fact that L is formally self-adjoint implies that μ_t is symmetric (that is, $d\mu_t(x^{-1}) = d\mu_t(x)$).

Let h be the distribution on $G \times (0,\infty)$ defined by μ_t:

$$\langle h, u \otimes v \rangle = \int_0^\infty \int_G u(x)v(t)d\mu_t(x)dt \qquad (u \in C_0^\infty(G),\ v \in C_0^\infty((0,\infty))).$$

Then we have

$$\langle h, Lu \otimes v \rangle = \int_0^\infty \int_G Lu(x)v(t)d\mu_t(x)dt = \int_0^\infty \int_G Lu(x)v(t)d\mu_t(x^{-1})dt$$

$$= \int_0^\infty (Lu*\mu_t)(0)v(t)dt = -\int_0^\infty \partial_t(u*\mu_t)(0)v(t)dt$$

$$= \int_0^\infty (u*\mu_t)(0)\partial_t v(t)dt = \int_0^\infty \int_G u(x)\partial_t v(t)d\mu_t(x)dt$$

$$= \langle h, u \otimes \partial_t v \rangle .$$

But this says that h is a distribution solution of $(\partial_t + L)h = 0$, so by the hypoellipticity of $\partial_t + L$, h is C^∞ on $G \times (0,\infty)$. Clearly $d\mu_t(x) = h(x,t)dx$, so properties (ii) and (iii) follow from the corresponding properties of μ_t. As for property (iv), we observe that since $L(u \circ \delta_r) \circ \delta_{1/r} = r^2 Lu$ for all $u \in C_0^\infty$, the family of measures associated to $r^2 L$ by Hunt's theorem — namely, $\{\mu_{r^2 t}\}_{t>0}$ — must be given by

$$u*\mu_{r^2 t} = [(u \circ \delta_r)*\mu_t] \circ \delta_{1/r}.$$

In other words,

$$\int u(xy^{-1})h(y,r^2 t)dy = \int u(x(ry^{-1}))h(y,t)dy = \int u(xy^{-1})h(r^{-1}y,t)r^{-Q}dy,$$

so that $h(y,r^2 t) = r^{-Q}h(r^{-1}y,t)$, which proves (iv). #

If f is a locally integrable function on G, let us define

(1.69) $$H_t f(x) = [f*h(\cdot,t)](x) = \int f(xy^{-1})h(y,t)dy,$$

provided that the integral exists. We call h the <u>heat kernel</u> for G and $\{H_t\}_{t>0}$ the <u>heat semigroup</u> for G.

(1.70) COROLLARY. (a) H_t <u>is a contraction operator on</u> L^p <u>for</u> $1 \leq p \leq \infty$ <u>and</u> $t > 0$.

(b) $\|H_t f - f\|_p \to 0$ <u>as</u> $t \to 0$ <u>if</u> $p < \infty$ <u>and</u> $f \in L^p$, <u>or if</u> $p = \infty$ <u>and</u> $f \in C_0$.

(c) $H_s H_t = H_{s+t}$ <u>for all</u> $s,t > 0$.

(d) <u>If</u> $f \in L^p$ $(1 \leq p \leq \infty)$, <u>the function</u> $u(x,t) = H_t f(x)$ <u>satisfies</u> $(\partial_t + L)u = 0$.

<u>Proof</u>: Since $\|h(\cdot,t)\|_1 = 1$, (a) follows from Young's inequality. To obtain (b), let $\phi(x) = h(x,1)$. Then $\int \phi = 1$ and $\phi_\varepsilon(x) = \varepsilon^{-Q}\phi(x/\varepsilon) = h(x,\varepsilon^2)$ so (b) follows from Proposition 1.20. (c) and (d) are obvious.

(1.71) PROPOSITION. <u>Extend</u> h to $G \times \mathbb{R}$ <u>by setting</u> $h(x,t) = 0$ <u>for</u> $t \leq 0$. <u>Then</u> h <u>is locally integrable on</u> $G \times \mathbb{R}$, and $(\partial_t + L)h = \delta$ <u>in the sense of distributions, where</u> δ <u>is the point mass at</u> $(0,0) \in G \times \mathbb{R}$.

<u>Proof</u>: It follows from Proposition 1.68 (ii) that h is integrable over $G \times (a,b)$ whenever $-\infty < a < b < \infty$. Given $\varepsilon > 0$, define $h^\varepsilon(x,t) = h(x,t)$ if $t > \varepsilon$ and $h^\varepsilon(x,t) = 0$ otherwise. Then $\|h^\varepsilon - h\|_1 = \varepsilon$, so it suffices to show that $\lim_{\varepsilon \to 0}(\partial_t + L)h^\varepsilon = \delta$, and this is equivalent to the assertion that for every $u \in C_0^\infty(G \times \mathbb{R})$, $(\partial_t + L)(u * h^\varepsilon)$ [convolution on $G \times \mathbb{R}$] converges pointwise to u as $\varepsilon \to 0$. But since $(\partial_t + L)h = 0$

for $t > 0$,

$$(\partial_t + L)(u*h^\varepsilon)(x,t) = (\partial_t + L)\int_{-\infty}^{t-\varepsilon}\int_G u(y,s)h(y^{-1}s,t-s)dyds$$

$$= \int_G u(y,t-\varepsilon)h(y^{-1}s,\varepsilon)dy$$

$$= \int_G [u(y,t-\varepsilon)-u(y,t)]h(y^{-1}x,\varepsilon)dy + \int_G u(y,t)h(y^{-1}x,\varepsilon)dy$$

$$= I_1^\varepsilon + I_2^\varepsilon.$$

Clearly $|I_1^\varepsilon| \leq \sup_{y,t} |u(y,t-\varepsilon)-u(y,t)| \to 0$ as $\varepsilon \to 0$. On the other hand $I_2^\varepsilon = H_\varepsilon u(\cdot,t) \to u(\cdot,t)$ as $\varepsilon \to 0$ by Corollary 1.70. #

(1.72) Corollary. h <u>is</u> C^∞ <u>on</u> $G \times \mathbb{R}\backslash\{(0,0)\}$.

<u>Proof</u>: This again follows from the hypoellipticity of $\partial_t + L$. #

We observe that if we regard $G \times \mathbb{R}$ as a homogeneous group with dilations $r(x,t) = (rx,r^2t)$, then h is homogeneous of degreee $-Q$. It follows immediately that for any $k \in \mathbb{N}$ and any multiindex I, $\partial_t^k X^I h$ is homogeneous of degree $-Q-d(I)-2k$, that is,

$$(1.73) \qquad \partial_t^k X^I h(rx,r^2t) = r^{-Q-d(I)-2k}\partial_t^k X^I h(x,t).$$

(1.74) PROPOSITION. $h(\cdot,t) \in S$ <u>for each</u> $t > 0$.

<u>Proof</u>. Since $h(\cdot,t) = 0$ for $t \leq 0$, it follows from Corollary 1.72 that $X^I h(x,t)$ vanishes to infinite order as $t \to 0$ whenever $x \neq 0$. More

precisely, for any $N \in \mathbb{N}$ and any multiindex I,

$$\sup_{|y|=1} |X^I h(y,s)| \leq C_{IN} s^N \quad \text{for} \quad 0 < s \leq 1.$$

But then for any $x \neq 0$ in G and $t > 0$, by (1.73) we have

$$|X^I h(x,t)| = |x|^{-Q-d(I)} |X^I h(|x|^{-1}x, |x|^{-2}t)|$$

$$\leq C_{IN} t^N |x|^{-Q-d(I)-2N},$$

which shows that $h(\cdot,t) \in \mathcal{S}$.

Remark. In view of this result, the operators H_t defined by (1.69) can be extended to act on $\mathcal{S}'$, and we have $H_t f \to f$ in $\mathcal{S}'$ as $t \to 0$ for all $f \in \mathcal{S}'$.

(1.75) PROPOSITION. *For each* $k \in \mathbb{N}$ *and each multiindex* I *there is a constant* $C > 0$ *such that for all* $t > 0$,

$$\int |\partial_t^k X^I h(x,t)| dx \leq C t^{-k-(d(I)/2)}.$$

Proof: When $|I| = k = 0$ this is just the fact that $\int h(x,t)dx = 1$. If $d(I)+k > 0$ we observe that by (1.73),

$$|\partial_t^k X^I h(x,t)| \leq C(\sqrt{t} + |x|)^{-Q-d(I)-2k}$$

where $C = \sup\{|\partial_t^k X^I h(y,s)| : \sqrt{s} + |y| = 1\}$. Therefore

$$\int |\partial_t^k X^I h(x,t)| dx \leq C \int_{|x| \leq \sqrt{t}} t^{-k-(Q+d(I))/2} dx + C \int_{|x| > \sqrt{t}} |x|^{-Q-d(I)-2k} dx$$

$$= C' t^{-k-(d(I)/2)}. \quad \#$$

Notes and References

Sections A through E: Some of this material is folklore, and some of it is derived from Knapp and Stein [1] and Folland [1]; see also Goodman [1]. The Taylor inequalities (1.37) and (1.42) and the results in Section E are new.

Section F: For the original theorems of Wiener and Whitney, see Stein [2]. The variants presented here, which are valid on arbitrary spaces of homogeneous type, are in Coifman and Weiss [1], [2]; see also Koranyi and Vagi [1].

Section G: These results are due to Folland [1].

CHAPTER 2

Maximal Functions and Atoms

Here we begin our development of the basic ideas of H^p theory. After reviewing some facts concerning maximal functions on L^p, $p \geq 1$, we turn to the "grand maximal function" in terms of which we define H^p. Atoms are also defined, and it is proved that "atomic H^p", namely $H^p_{q,a}$ is contained in H^p.

We shall be working on a fixed homogeneous group G, and we recall that if ϕ is a function on G and $t > 0$, we set $\phi_t = t^{-Q}\phi \circ \delta_{1/t}$; this notation will be used throughout.

If $f \in \mathcal{S}'$ and $\phi \in \mathcal{S}$, we define the <u>nontangential maximal function</u> $M_\phi f$ and the <u>radial maximal function</u> $M^0_\phi f$ of f with respect to ϕ by

(2.1) $$M_\phi f(x) = \sup \{|f*\phi_t(y)| : |x^{-1}y| < t,\ 0 < t < \infty\},$$

(2.2) $$M^0_\phi f(x) = \sup_{0<t<\infty} |f*\phi_t(x)|.$$

The same definitions will apply if f and ϕ are any two distributions such that $(x,t) \to f*\phi_t(x)$ is a continuous function $G \times (0,\infty)$ — for example, if $f \in L^p$ and $\phi \in L^q$ where $1 \leq p \leq \infty$ and $p^{-1} + q^{-1} = 1$. Clearly $M^0_\phi f \leq M_\phi f$ pointwise for all f, ϕ. We shall usually write the right hand side of (2.1) more briefly as

$$\sup_{|x^{-1}y|<t<\infty} |f*\phi_t(y)|.$$

(2.3) PROPOSITION. $M_\phi f$ and $M_\phi^0 f$ <u>are measurable functions from</u> G <u>to</u> $[0,\infty]$. <u>Moreover</u> $M_\phi f$ <u>is lower semicontinous</u>.

<u>Proof</u>: If $M_\phi f(x) > \alpha$, there exist $y \in G$, $t > 0$ such that $|x^{-1}y| < t$ and $|f*\phi_t(y)| > \alpha$. But then if $|x^{-1}z| < \gamma^{-1}(t-|x^{-1}y|)$ we have $|z^{-1}y| < t$ and hence $M_\phi f(z) > \alpha$. Thus $M_\phi f$ is lower semicontinuous, hence measurable. Since $(x,t) \to f*\phi_t(x)$ is continuous on $G \times (0,\infty)$, the right hand side of 2.2) is unchanged if we restrict t to be rational, whence $M_\phi^0 f$ is measurable. #

<u>Remark</u>: We shall also consider "grand maximal functions" of the form $Mf = \sup_{\phi\in A} M_\phi f$ or $M^0 f = \sup_{\phi\in A} M_\phi^0 f$ where A is some suitable class of functions. It follows from Proposition 2.3 that Mf is always lower semi-continous. Moreover, the family A will always have a countable dense subset in a suitable topology, from which it will follow that $M^0 f$ is measurable. We shall not comment further on this point. See also Proposition 2.8.

The starting point of H^p theory is the following version of the classical maximal theorem:

(2.4) THE MAXIMAL THEOREM. <u>For</u> $\lambda > Q$, <u>let</u> A_λ <u>be the set of measurable functions</u> ϕ <u>on</u> G <u>such that</u> $|\phi(x)| \leq (1+|x|)^{-\lambda}$. (<u>Thus</u> $A_\lambda \subset L^q$ <u>for</u> $1 \leq q \leq \infty$.) <u>If</u> $f \in L^p$ $(1 \leq p \leq \infty)$, <u>let</u>

$$M^{(\lambda)}f(x) = \sup_{\phi\in A_\lambda} M_\phi f(x).$$

<u>Then for each</u> $\lambda > Q$ <u>there exist constants</u> C_λ, C'_λ <u>such that</u>

(a) $|\{x : M^{(\lambda)}f(x) > \alpha\}| \leq C_\lambda \|f\|_1/\alpha$ <u>for all</u> $f \in L^1$ <u>and</u> $\alpha > 0$,

(b) $\|M^{(\lambda)}f\|_p \leq C'_\lambda p(p-1)^{-1}\|f\|_p$ for all $f \in L^p$, $1 < p \leq \infty$.

Proof: First we prove (a). We fix $\lambda > Q$ and write Mf instead of $M^{(\lambda)}f$. Given $f \in L^1$, $\alpha > 0$, and $R > 0$, let

$$E_\alpha^R = \{x : Mf(x) > \alpha \text{ and } |x| < R\}.$$

Then for each $x \in E_\alpha^R$ we can pick $y = y(x)$, $t = t(x)$, and $\phi \in A_\lambda$ such that $|x^{-1}y| < t$ and $|f*\phi_t(y)| > \alpha$. Thus

$$\alpha < \int |f(z)||\phi_t(z^{-1}y)|dz = t^{-Q}\int |f(z)||\phi((z^{-1}y)/t)|dz.$$

We write the last integral as a sum of integrals over the regions $|z^{-1}y| < t$ and $2^k t \leq |z^{-1}y| < 2^{k+1}t$ $(k = 0,1,2\ldots)$. Since $\phi \in A_\lambda$ we obtain

$$t^Q\alpha < \int_{|z^{-1}y|<t} |f(z)|dz + \sum_0^\infty 2^{-k\lambda}\int_{2^k t\leq|z^{-1}y|<2^k t} |f(z)|dz$$

$$\leq \int_{B(t,y)} |f(z)|dz + \sum_0^\infty 2^{-k\lambda}\int_{B(2^{k+1}t,y)} |f(z)|dz$$

$$\leq (1 + \sum_0^\infty 2^{-k(\lambda-Q)})\sup_{k\geq 0} 2^{-(k-1)Q}\int_{B(2^k t,y)} |f(z)|dz.$$

Hence if we set

$$A = A(\lambda) = 2^Q(1 + \sum_0^\infty 2^{-k(\lambda-Q)})$$

(which is finite since $\lambda > Q$), for some $k = k(x)$ we have

$$(2^k t)^{-Q}\int_{B(2^k t,y)} |f(z)|dz > \alpha/A.$$

But $|x^{-1}y| < t$, so $B(2^k t, y) \subset B(\gamma 2^{k+1}t, x)$, whence

$$\int_{B(\gamma 2^{k+1}t,x)} |f(z)|dz > \frac{\alpha}{A}(2^k t)^Q = \frac{\alpha}{A(2\gamma)^Q}|B(\gamma 2^{k+1}t,x)|.$$

In other words, for each $x \in E_\alpha^R$ there is a ball $B(r(x),x)$ (where $r(x) = \gamma 2^{k(x)+1}t(x)$) such that, with $A' = (2\gamma)^Q A$,

$$\int_{B(r(x),x)} |f(z)|dz \geq \frac{\alpha}{A'}|B(r(x),x)|.$$

By the Wiener lemma (1.66), we can choose a sequence $\{x_j\}$ in E_α^R so that the balls $B(r(x_j),x_j)$ are disjoint and the balls $B(4\gamma r(x_j),x_j)$ cover E_α^R. Then

$$|E_\alpha^R| \leq \Sigma_j |B(4\gamma r(x_j),x_j)| \leq \frac{(4\gamma)^Q A'}{\alpha} \Sigma_j \int_{B(r(x_j),x_j)} |f(z)|dz$$

$$\leq \frac{(4\gamma)^Q A'}{\alpha} \|f\|_1.$$

Since A' is independent of R, we obtain assertion (a) by letting $R \to \infty$.

Assertion (b) is trivial for $p = \infty$: in fact,

$$\|Mf\|_\infty \leq A_1 \|f\|_\infty \quad \text{where} \quad A_1 = \int (1+|x|)^{-\lambda} dx.$$

The result for $1 < p < \infty$ then follows from the Marcinkiewicz interpolation theorem, but this situation is simple enough that we can give a direct proof. Let A', A_1 be as above and set $A_0 = (4\gamma)^Q A'$. Given $f \in L^p$ $(1 < p < \infty)$ and $\alpha > 0$, set $g(x) = f(x)$ if $|f(x)| \geq \alpha/2A_1$ and $g(x) = 0$ otherwise.

Then $|f| \leq |g| + \alpha/2A_1$, whence $Mf \leq Mg + \alpha/2$, whence

$$\{x : Mf(x) > \alpha\} \subset \{x : Mg(x) > \alpha/2\}$$

and thus by part (a), if $\lambda(\alpha) = |\{x : Mf(x) > \alpha\}|$,

$$\lambda(\alpha) \leq \frac{2A_0}{\alpha} \|g\|_1 = \frac{2A_0}{\alpha} \int_{|f|>\alpha/2A_1} |f(z)|dz.$$

Therefore

$$\int Mf(x)^p dx = p \int_0^\infty \alpha^{p-1} \lambda(\alpha) d\alpha$$

$$\leq p \int_0^\infty \alpha^{p-1} [2A_0\alpha^{-1} \int_{|f|>\alpha/2A_1} |f(z)|dz] d\alpha$$

$$= 2A_0 p \int_G \int_0^{2A_1|f(z)|} \alpha^{p-2} d\alpha |f(z)|dz$$

$$= \frac{2A_0(2A_1)^{p-1}p}{p-1} \int_G |f(z)|^p dz,$$

from which the desired result is immediate. #

(2.5) COROLLARY. *If* ϕ *is a measurable function on* G *such that* $|\phi(x)| \leq A(1+|x|)^{-\lambda}$ *for some* $A > 0$ *and* $\lambda > Q$, *then*:

(a) $|\{x : M_\phi f(x) > \alpha\}| \leq AC_\lambda \|f\|_1/\alpha$ *for all* $f \in L^1$ *and* $\alpha > 0$,

(b) $\|M_\phi f\|_p \leq AC'_\lambda p(p-1)^{-1} \|f\|_p$ *for all* $f \in L^p$, $1 < p < \infty$.

An important special case of this is the Hardy-Littlewood maximal function, which is obtained by taking ϕ to be the characteristic function of the unit ball $B(1,0)$. We shall denote it by M_{HL}:

$$M_{HL}f(x) = \sup\{|B|^{-1}|\int_B f(y)dy| : B \text{ is a ball containing } x\}.$$

One of the traditional applications of the maximal theorem is the following complement to Proposition 1.20. Some notation: if u is a continuous function on $G \times (0,\infty)$, we shall interpret the equation

$$\lim_{|x^{-1}y| < t \to 0} u(y,t) = A$$

to mean that $u(y,t) \to A$ as (y,t) approaches $(x,0)$ in the set $\{(y,t) : |x^{-1}y| < t < \infty\}$.

(2.6) THEOREM. *Suppose* ϕ *satisfies the hypotheses of Corollary* 2.5, *and let* $a = \int\phi$. *Then if* $f \in L^p$, $1 \leq p \leq \infty$,

$$\lim_{|x^{-1}y| < t \to 0} f*\phi_t(y) = af(x) \quad \textit{for almost every } x \in G.$$

Proof: First suppose $p = 1$. Given $\varepsilon > 0$, choose $g \in C_c(G)$ such that $\|g-f\|_1 < \varepsilon$. Then $\lim_{|x^{-1}y| < t \to 0} g*\phi_t(x) = ag(x)$ for every x by Proposition 1.20, so

$$\limsup_{|x^{-1}y| < t \to 0} |f*\phi_t(y) - af(x)|$$

$$\leq \sup_{|x^{-1}y| < t < \infty} |f*\phi_t(y) - g*\phi_t(y)| + |a||f(x) - g(x)|$$

$$\leq M_\phi(f-g)(x) + |a||f(x) - g(x)|.$$

Hence for any $\alpha > 0$, by Corollary 2.5,

$$|\{x : \limsup_{|x^{-1}y| < t \to 0} |f*\phi_t(y) - af(x)| > \alpha\}|$$

$$\leq |\{x : M_\phi(f-g)(x) > \alpha/2\}| + |\{x : |a||f(x) - g(x)| > \alpha/2\}|$$

$$\leq C\|f-g\|_1/\alpha < C\varepsilon/\alpha.$$

Since ε is arbitrary,

$$|\{x : \limsup_{|x^{-1}y| < t \to 0} |f*\phi_t(y) - af(x)| > \alpha\}| = 0$$

for every $\alpha > 0$, so we are done.

If $p > 1$, it suffices to show that for each $r > 0$, $\lim_{|x^{-1}y| < t \to 0} f*\phi_t(y) = af(x)$ for almost every $x \in B(r,0)$. Let $f_r = f\chi_{B(r,0)}$. Then $f_r \in L^1$, so $\lim_{|x^{-1}y| < t \to 0} f_r*\phi_t(y) = f_r(x) = f(x)$ for almost every $x \in B(r,0)$. On the other hand, if $x \in B(r,0)$ we can choose $\varepsilon > 0$ so small that $B(2\gamma\varepsilon,x) \subset B(r,0)$. Then if $y \in B(t,x)$ with $t < \varepsilon$ we have $B(\varepsilon,y) \subset B(r,0)$, so if $p^{-1} + q^{-1} = 1$,

$$\begin{aligned} |(f-f_r)*\phi_t(y)| &= \left|\int_{|z|\geq r} f(z)\phi_t(z^{-1}y)dz\right| \\ &\leq \|f\|_p\left(\int_{|z|\geq r} |\phi_t(z^{-1}y)|^q dz\right)^{1/q} \\ &\leq \|f\|_p\left(\int_{|w|\geq \varepsilon} |\phi_t(w)|^q dw\right)^{1/q} \\ &\leq A\|f\|_p t^{-Q}\left(\int_{|w|\geq \varepsilon} |w/t|^{-\lambda q} dw\right)^{1/q} \\ &= Ct^{\lambda-Q} \to 0 \text{ as } t \to 0. \quad \# \end{aligned}$$

The converse of the maximal theorem (for $p > 1$) is also true. In fact, we have the following stronger result.

(2.7) THEOREM. <u>Suppose</u> $1 \leq p \leq \infty$ <u>and</u> $f \in S'$. <u>If there exists</u> $\phi \in S$ <u>with</u> $\int \phi = 1$ <u>such that</u> $M_\phi^0 f \in L^p$, <u>then</u> $f \in L^p$.

<u>Proof</u>: By Proposition 1.49, $f*\phi_t \to f$ in S' as $t \to 0$. But since $M_\phi^0 f \in L^p$, the set $\{f*\phi_t : t > 0\}$ is bounded in L^p, so if $p > 1$ there is a sequence $t_j \to 0$ such that $f*\phi_{t_j}$ converges in the weak * topology of L^p. Hence $f = \lim f*\phi_{t_j} \in L^p$. Similarly, if $p = 1$ there is a sequence $t_j \to 0$ such that $f*\phi_{t_j}$ converges in the weak * topology of measures to a measure μ, and we claim that μ is absolutely continuous. Since $M_\phi^0 f \in L^1$, for any $\varepsilon > 0$ there exists $\delta > 0$ such that $\int_E M_\phi^0 f < \varepsilon$ whenever $|E| < \delta$. Given $N \subset G$ with $|N| = 0$, choose an open $U \subset N$ with $|U| < \delta$. Then for any $g \in C_c(U)$,

$$\left|\int g d\mu\right| = \lim \left|\int g(x) f*\phi_{t_j}(x)dx\right| \leq \|g\|_\infty \int_U M_\phi^0 f(x)dx < \|g\|_\infty \varepsilon .$$

But $|\mu|(U) = \sup\{\left|\int g d\mu\right| : g \in C_c(U), \|g\|_\infty \leq 1\}$, so $|\mu|(U) < \varepsilon$. Since ε is arbitrary, $|\mu|(N) = 0$, and we are done. #

It follows from Theorems 2.4 and 2.7 that for $p > 1$, a tempered distribution f is in L^p if and only if $M_\phi f \in L^p$ for every $\phi \in S$; moreover, in this case $M_\phi f$ is p-th power integrable uniformly in ϕ as ϕ ranges over the sets A_λ of Theorem 2.4. Our aim now is to investigate distributions f which satisfy analogous conditions for $p \leq 1$, that is, $M_\phi f \in L^p$ for all $\phi \in S$, with some uniformity in ϕ. To make this precise, we introduce the following definitions.

If $N \in \mathbb{N}$ and $f \in S'$, we define the nontangential and radial grand maximal functions $M_{(N)}f$ and $M^0_{(N)}f$ by

$$M_{(N)}f(x) = \sup_{\phi \in S, \|\phi\|_{(N)} \leq 1} M_\phi f(x),$$

$$M^0_{(N)}f(x) = \sup_{\phi \in S, \|\phi\|_{(N)} \leq 1} M^0_\phi f(x).$$

(Here $\|\phi\|_{(N)}$ is the norm introduced in Chapter 1, Section D.) Whether we use $M_{(N)}$ or $M^0_{(N)}$ is purely a matter of convenience, as the following proposition shows.

(2.8) PROPOSITION. *There is a constant* C *depending only on* N *such that for all* $f \in S'$ *and* $x \in G$,

$$M^0_{(N)}f(x) \leq M_{(N)}f(x) \leq CM^0_{(N)}f(x).$$

Proof: The first inequality is obvious. As for the second, observe that $f*\phi_t(y) = f*\psi_t(x)$ where $\psi(z) = \phi(z\cdot\delta_{1/t}(x^{-1}y))$, so

$$M_{(N)}f(x) = \sup\{M^0_\psi f(x) : \psi(z) = \phi(zw) \text{ for some } \phi \in S \text{ with } \|\phi\|_{(N)} \leq 1 \text{ and some } w \in G \text{ with } |w| < 1\}.$$

But by Proposition 1.46, there exists $C = C_N$ such that if $\psi(z) = \phi(zw)$ and $|w| < 1$ then $\|\psi\|_{(N)} \leq C\|\phi\|_{(N)}$. Hence

$$M_{(N)}f(x) \leq \sup_{\psi \in S, \|\psi\|_{(N)} \leq C} M^0_\psi f(x) = CM^0_{(N)}f(x). \quad \#$$

If $f \in L^p$ and $p > 1$, then $M_{(N)}f \in L^p$ for all $N \in \mathbb{N}$. That the condition $f \in L^p$ is inappropriate to obtain $M_{(N)}f \in L^p$ when $p \leq 1$ can be seen by considering the characteristic function χ of the unit ball $B(1,0)$. Certainly $\chi \in L^p$ for all $p > 0$. However, suppose $\phi \in \mathcal{S}$ is nonnegative and satisfies $\phi(0) = 1$. Then there is a ball $B(r,0)$ on which $\phi \geq 1/2$. If $x \in G$ and $t = \gamma(1+|x|)/r$ we have $B(1,x) \subset B(rt,0)$, so

$$\chi * \phi_t(x) = \int_{B(1,x)} \phi_t(y)dy \geq t^{-Q}/2 = C(1+|x|)^{-Q}.$$

From this it is clear that $M_\phi \chi \in L^p$ only when $p > 1$. However, for any $p > 0$ we can exhibit large classes of distributions whose grand maximal functions are in L^p. The problem is to ensure that the maximal functions vanish sufficiently rapidly at infinity, and this can be accomplished by assuming some vanishing moment conditions. We proceed to the formal definitions.

If $0 < p \leq 1$ and $a \in \Delta$, we shall say that a is p-<u>admissible</u> if $a \geq \max\{a' \in \Delta : a' \leq Q(p^{-1}-1)\}$. (If $\Delta = \mathbb{N}$, this means that $a \geq [Q(p^{-1}-1)]$.)* If in addition $1 \leq q \leq \infty$, we shall call the ordered triplet (p,q,a) <u>admissible</u> if a is p-admissible and $p < q$. (The latter condition excludes only the case $p = q = 1$.)

Suppose (p,q,a) is admissible. A (p,q,a)-<u>atom</u> is a compactly supported L^q function f such that

(i) there is a ball B whose closure contains $\operatorname{supp}(f)$ such that $\|f\|_q \leq |B|^{(1/q)-(1/p)}$,

---

(*) Recall that Δ is a set of real numbers generated by the exponents of homogeneity. See the definition on p. 21.

(ii) $\int f(x)P(x)dx = 0$ for all $P \in P_a$.

If B is any ball satisfying condition (i), we shall say that f is _associated_ to B.

Condition (ii) in the definition of atom is the essential one. Condition (i) merely imposes a normalization on f whose utility will become apparent later; it implies that $f \in L^p$ and that

$$\int |f|^p \leq (\int |f|^q)^{p/q} |B|^{1-(p/q)} \leq 1.$$

Finally, if $p > 0$ we set

$$N_p = \min\{N \in \mathbb{N} : N \geq \min\{b \in \Delta : b > Q(p^{-1}-1)\}\}.$$

Thus if $p > 1$ we have $N_p = 0$, while if $p \leq 1$, $N_p = [a]+1$ where a is the smallest element of Δ which is p-admissible. (If $\Delta = \mathbb{N}$ and $p \leq 1$, $N_p = [Q(p^{-1}-1)]+1$.)

The point of all these definitions appears in the following theorem.

(2.9) THEOREM. _If_ (p,q,a) _is admissible and_ $N \geq N_p$, _there is a constant_ $C = C(p,q,a,N) < \infty$ _such that_ $\|M_{(N)}f\|_p \leq C$ _for all_ (p,q,a)-_atoms_ f.

Proof: Lef f be a (p,q,a)-atom associated to $B = B(r,x_0)$. Since all the relevant definitions are invariant under left translations, we may assume without loss of generality that $x_0 = 0$. Let $\tilde{B} = B(2\gamma\beta^N r,0)$. We shall estimate $M_{(N)}f$ on $\tilde{B}$ and $\tilde{B}^c$ separately.

On $\tilde{B}$ we use the maximal theorem (2.4). If $q > 1$ we have

$$\int_{\tilde{B}} M_{(N)}f(x)^p dx \leq \left(\int M_{(N)}f(x)^q dx\right)^{p/q} |\tilde{B}|^{1-(p/q)} \tag{2.10}$$

$$\leq C\|f\|_q^p |B|^{1-(p/q)} = C.$$

If $q = 1$, let $E_\alpha = \{x : M_{(N)}f(x) > \alpha\}$. Then for all $\alpha > 0$,

$$|E_\alpha \cap \tilde{B}| \leq \min\{|E_\alpha|, |\tilde{B}|\}$$

$$\leq \min\{C\|f\|_1/\alpha, |\tilde{B}|\}$$

$$= \min\{C|B|^{1-(1/p)}/\alpha, (2\gamma\beta^N)^Q|B|\}$$

$$\leq C' \min\{|B|^{1-(1/p)}/\alpha, |B|\}.$$

The last two quantities in curly brackets are equal when $\alpha = |B|^{-1/p}$. Thus, since $p < q = 1$,

$$\int_{\tilde{B}} M_{(N)}f(x)^p dx = \int_0^\infty p\alpha^{p-1}|E_\alpha \cap \tilde{B}| d\alpha \tag{2.11}$$

$$\leq C'\left[\int_0^{|B|^{-1/p}} |B| p\alpha^{p-1} d\alpha + \int_{|B|^{-1/p}}^\infty |B|^{1-(1/p)} p\alpha^{p-2} d\alpha\right]$$

$$= C''.$$

Next, suppose $\phi \in S$ and $\|\phi\|_{(N)} \leq 1$. Let $b = \min\{b' \in \Delta : b' > Q(p^{-1}-1)\}$ and $c = \max\{c' \in \Delta : c' < b\}$, and for each $x \in G$ let P_x be the right Taylor polynomial of ϕ at x of homogeneous degree c. Then by Theorem 1.50 there is a constant $C > 0$, independent of ϕ, such that whenever $|x| \geq 2\gamma\beta^N|y|$,

$$(2.12) \qquad |\phi(y^{-1}x) - P_x(y^{-1})| \leq C|y|^b|x|^{-Q-b}.$$

Therefore, if $y \in B$ and $x \in \tilde{B}^c$,

$$|\phi((y^{-1}x)/t) - P_{x/t}(y^{-1}/t)| \leq Ct^Q|y|^b|x|^{-Q-b}.$$

Since a is p-admissible we have $a \geq c$, and hence if q' is the conjugate to q,

$$\begin{aligned} |f*\phi_t(x)| &= |t^{-Q} \int f(y)[\phi((y^{-1}x)/t) - P_{x/t}(y^{-1}/t)]dy| \\ &\leq C|x|^{-Q-b} \int |f(y)||y|^b dy \\ &\leq C|x|^{-Q-b} \left(\int |f(y)|^q dy\right)^{1/q} \left(\int_B |y|^{bq'} dy\right)^{1/q'} \\ &\leq C|x|^{-Q-b} |B|^{(b/Q)-(1/p)+1}. \end{aligned}$$

This being true for all $x \in \tilde{B}^c$ and all $\phi \in \mathcal{S}$ with $\|\phi\|_{(N)} \leq 1$, by Proposition 2.8 we have

$$\int_{\tilde{B}^c} M_{(N)}f(x)^p dx \leq C|B|^{(bp/Q)-1+p} \int_{\tilde{B}^c} |x|^{(-Q-b)p} dx.$$

But $b > Q(p^{-1}-1)$, so $(-Q-b)p < -Q$, and hence

$$(2.13) \qquad \int_{\tilde{B}^c} M_{(N)}f(x)^p dx \leq C|B|^{(bp/Q)-1+p} |\tilde{B}|^{1-(Q+b)p/Q} = C'.$$

Combining (2.13) with (2.10) or (2.11), we are done. #

After these preliminaries we now make the following definitions. If $0 < p \leq \infty$, we define the *Hardy space* H^p to be

$$H^p = \{f \in S' : M_{(N_p)} f \in L^p\}.$$

If (p,q,a) is an admissible triplet, we define the *atomic Hardy space* $H^p_{q,a}$ to be the set of all tempered distributions of the form $\Sigma_1^\infty \lambda_i f_i$ (the sum converging in the topology of S') where each f_i is a (p,q,a)-atom, $\lambda_i \geq 0$, and $\Sigma_1^\infty \lambda_i^p < \infty$.

Several remarks are in order concerning these definitions.

(1) If $p > 1$ then $H^p = L^p$, and $H^1 \subset L^1$, by Theorems 2.4 and 2.7.

(2) The condition $\|\phi\|_{(N_p)} \leq 1$ in the definition of $M_{(N_p)}f$ is essentially the weakest one which allows the arguments in the proof of Theorem 2.9 to be carried out. One needs estimates for the derivatives of ϕ of order $\leq N_p$ in order to obtain the estimate (2.12). (If the group G is stratified, one can do a bit better: one only needs estimates for the derivatives of ϕ of homogeneous degree $\leq N_p$, since then one can use the stratified Taylor inequality (1.42) to obtain (2.12).) However, the estimates on these derivatives implied by the condition $\|\phi\|_{(N_p)} \leq 1$ are only roughly optimal. For example, if $p > 1$, the condition $\|\phi\|_{(0)} = \|\phi\|_{(N_p)} \leq 1$ means that $|\phi(x)| \leq (1+|x|)^{-(Q+1)}$, and one can replace the number $Q+1$ by any $\lambda > Q$ as in Theorem 2.4. We shall not pursue here the point of precise decay conditions on ϕ. We only remark that in some problems the optimal conditions on ϕ (both of decay and of smoothness) have some interest; for this see Stein, Taibleson, and Weiss [1].

(3) For any $N \in \mathbb{N}$ we could define

$$H^p_{(N)} = \{f \in S' : M_{(N)} f \in L^p\}. \tag{2.14}$$

Clearly $H^p_{(N)} \subset H^p_{(M)}$ whenever $N \leq M$. If $p > 1$ it follows from Theorems 2.4 and 2.7 that $H^p_{(N)} = L^p$ for all $N \geq 0 = N_p$. Moreover, it will follow from the results of Chapter 3 that $H^p_{(N)} = H^p$ for all $N \geq N_p$ even when $p \leq 1$. On the other hand, we shall also show in Chapter 3 that $H^p_{q,a} = H^p$ whenever (p,q,a) is admissible. Bearing this in mind, one sees from the preceding remark that the theory breaks down when $N < N_p$. The moral is that one needs <u>some</u> control over the functions $\phi \in S$ which enter into the grand maximal function, and the smaller p is, the more control one needs.

(4) If $f \in H^p_{q,a}$, the representation $f = \Sigma_1^\infty \lambda_i f_i$ of f as a linear combination of atoms is far from unique. Any such representation will be called an <u>atomic decomposition</u> of f.

We now define topologies on H^p and $H^p_{q,a}$ for $p \leq 1$. (Of course, if $p > 1$ we use the L^p norm on H^p.) Namely, we define the quasi-norms ρ^p and $\rho^p_{q,a}$ by

$$\rho^p(f) = \|M_{(N_p)} f\|_p^p,$$

$$\rho^p_{q,a}(f) = \inf\{\Sigma \lambda_i^p : \Sigma \lambda_i f_i \text{ is an atomic decomposition of } f \text{ into } (p,q,a)\text{-atoms}\}.$$

It is easily verified that the maps $(f,g) \to \rho^p(f-g)$ and $(f,g) \to \rho^p_{q,a}(f-g)$ are metrics on H^p and $H^p_{q,a}$ respectively, which make them into topological vector spaces (not locally convex, unless $p = 1$). The only slightly nontrivial point is that $\rho^p_{q,a}(f) = 0$ implies $f = 0$, but this follows from the next proposition.

(2.15) PROPOSITION. _If_ (p,q,a) _is admissible then_ $H^p_{q,a} \subset H^p \subset S'$, _the inclusions being continuous._

Proof: Let $C = C(p,q,a,N_p)$ be as in Theorem 2.9. If $f = \Sigma\lambda_i f_i \in H^p_{q,a}$ we have (since $p \leq 1$)

$$\rho^p(f) = \int [M_{(N_p)}(\Sigma\lambda_i f_i)]^p \leq \Sigma\lambda_i^p \int [M_{(N_p)} f_i]^p \leq C\Sigma\lambda_i^p .$$

Hence $\rho^p \leq C\rho^p_{q,a}$, so that $H^p_{q,a} \subset H^p$ continuously. Next, if $f \in H^p$ and $\phi \in S$, let $\tilde{\phi}(x) = \phi(x^{-1})$. Then

$$\left|\int f(x)\phi(x)dx\right| = |f*\tilde{\phi}(0)| \leq M_{\tilde{\phi}}f(y) \qquad \text{for } |y| < 1,$$

and hence

$$\left|\int f(x)\phi(x)dx\right|^p \leq \int_{|y|\leq 1} M_{\tilde{\phi}}f(y)^p dy \leq \|\tilde{\phi}\|_{(N_p)} \int_{|y|<1} M_{(N_p)}f(y)dy$$

$$\leq C_\phi \rho^p(f).$$

From this it follows that if $f_n \to f$ in H^p then $\int f_n\phi \to \int f\phi$ for all $\phi \in S$, so the inclusion $H^p \subset S'$ is continuous. #

<u>Remark</u>: The same proof shows that if $H^p_{(N)}$ is defined by (2.14) and given the obvious topology, the inclusion $H^p_{(N)} \subset S'$ is continuous, for all $N \in \mathbb{N}$.

(2.16) PROPOSITION. H^p <u>is complete</u>.

<u>Proof</u>: We need only consider $p \leq 1$, and it suffices to show that if $\{f_j\}$ is a sequence in H^p such that $\Sigma\rho^p(f_j) < \infty$, the series Σf_j converges in H^p. However, the partial sums of this series are Cauchy in H^p, hence in S' by Proposition 2.15, so the series Σf_j converges in S' to a distribution f. We have

$$[M_{(N_p)}f]^p \leq [\Sigma M_{(N_p)}f_j]^p \leq \Sigma[M_{(N_p)}f_j]^p,$$

hence $\rho^p(f) \leq \Sigma\rho^p(f_j) < \infty$ and $f \in H^p$. Similarly,

$$\rho^p(f - \Sigma_1^N f_j) \leq \sum_{N+1}^{\infty} \rho^p(f_j) \to 0 \quad \text{as} \quad N \to \infty,$$

so the series converges in H^p. #

We conclude with two more remarks about atomic H^p spaces.

(2.17) If f is a (p,q,a)-atom then $\rho^p_{q,a}(f)$ may be strictly less than one, since there may be more "efficient" atomic decompositions of f than f itself. However, for any $\varepsilon > 0$ there exist (p,q,a)-atoms f such that $\rho^p_{q,a}(f) > 1-\varepsilon$. If not, then for any (p,q,a)-atom f we could write $f = \Sigma\lambda_i f_i$ with $\Sigma\lambda_i^p < 1-(\varepsilon/2)$. Decomposing each f_i similarly, we would obtain $\rho^p_{q,a}(f) < (1-(\varepsilon/2))^2$, and by induction, $\rho^p_{q,a}(f) < (1-(\varepsilon/2))^j$, for all j, which would imply $f = 0$.

(2.18) If $f = \Sigma_1^N \lambda_i f_i$ is a finite linear combination of (p,q,a)-atoms, then

$$\int |f|^p \leq \Sigma_1^N \int \lambda_i^p |f_i|^p \leq \Sigma_1^N \lambda_i^p, \qquad \text{hence} \quad \int |f|^p \leq \rho_{q,a}^p(f).$$

Thus the identity map of finite linear combinations of atoms extends to a continuous linear map from $H_{q,a}^p$ to L^p. This map is <u>not</u> injective if $p < 1$. Its nullspace consists of distributions which are supported on sets of measure zero. For example, if x_1 and x_2 are distinct points of G, the difference $\delta_{x_1} - \delta_{x_2}$ of the point masses at x_1 and x_2 lies in $H_{\infty,0}^p$ for $Q/(Q+1) < p < 1$. We leave it as an exercise for the reader to construct an explicit atomic decomposition $\Sigma\lambda_i f_i$ of $\delta_{x_1} - \delta_{x_2}$ and to check that the series $\Sigma\lambda_i f_i$ converges to zero in L^p. (Hint: express $\delta_{x_1} - \delta_{x_2}$ as the limit of a suitable sequence of functions g_i, and then write $\lim g_i = g_1 + \Sigma_1^\infty (g_{i+1} - g_i)$.)

Notes and References

Theorems 2.4, 2.6, and 2.7 are routine adaptations of well-known results on $\mathbb{R}^n$; see e.g. Stein [2]. The characterization of classical H^p spaces of harmonic functions in terms of grand maximal functions is due to Fefferman and Stein [1]; see also Goldberg [1] for a localized version of this result and Geller [2] for an analogue on the Heisenberg group, as well as Chapter 8 of this monograph. H^p spaces defined in terms of maximal functions have been studied on $\mathbb{R}^n$ with non-isotropic dilations by Calderón and Torchinsky [1],[2] and (for p near 1) on spaces of homogeneous type by Macías and Segovia [2]; see also Mauceri, Picardello, and Ricci [1]. References concerning atomic H^p spaces will be given at the end of Chapter 3.

CHAPTER 3

Decomposition and Interpolation Theorems

The purpose of this chapter is twofold. The first main objective is to find an atomic decomposition for every element of H^p, $p \leq 1$, thereby completing the identification of H^p with "atomic H^p". The second is to present several "real" interpolation theorems for H^p spaces.

The principal tool used to reach both of these objectives is a refined version of the Calderón-Zygmund decomposition. Since matters here are complicated, it may be worthwhile to make a brief comparison with the more standard decomposition, which can be found in e.g. Stein [2], chapters 1 and 2. There we start with a function $f \in L^1$, and at an altitude α we break it up in terms of its maximal function as $f = g+b$, where the good function g can be controlled in any L^p norm, $1 \leq p \leq \infty$; the bad function b has only L^1 control, but has the redeeming feature of having zero integral on appropriate cubes which cover its support. Here we break up our distributions f in terms of the grand maximal function of f. The corresponding controls on b and g (given by Theorems 3.17 and 3.20) are in terms of their grand maximal functions. The extra devices need to achieve this are: (i), to make the break-up in terms of a smooth partition of unity, and (ii), to insure that sufficiently many moments vanish (and not just the zero moments).

A. The Calderón-Zygmund Decomposition

In this section we shall be working in the following setting. N will be a fixed nonnegative integer, and we abbreviate the maximal operators $M_{(N)}$, $M^0_{(N)}$ by M, M^0. f will be a tempered distribution such that

$$|\{x : Mf(x) > \alpha\}| < \infty \quad \text{for all} \quad \alpha > 0.$$

We shall also fix $\alpha > 0$ and set

$$\Omega = \{x : Mf(x) > \alpha\}.$$

Finally, we introduce the following three constants:

$$T_1 = 9\gamma\beta^N, \quad T_2 = 2\gamma^2 T_1 = 18\gamma^3\beta^N, \quad T_3 = 3\gamma T_2 = 54\gamma^4\beta^N.$$

(T_1, T_2, and T_3 are "all-purpose constants" which will intervene in various ways throughout this chapter; in individual situations they could sometimes be replaced by smaller constants. Notice that in $\mathbb{R}^n$ with the usual dilations $T_1 = 9$, $T_2 = 18$, and $T_3 = 54$.)

Since Ω is open (by Proposition 2.3) and $|\Omega| < \infty$, by the Whitney lemma (1.67) there exist $x_1, x_2, \ldots$ in Ω and $r_1, r_2, \ldots > 0$ such that:

(3.1) $\quad \Omega = \bigcup_1^\infty B(r_j, x_j)$.

(3.2) $\quad$ The balls $B(r_j/4\gamma, x_j)$ are disjoint.

(3.3) $\quad B(T_2 r_j, x_j) \cap \Omega^c = \emptyset$, but $B(T_3 r_j, x_j) \cap \Omega^c \neq \emptyset$.

(3.4) $\quad$ There exists $L \in \mathbb{N}$ (depending only on the group G) such that no point of Ω lies in more than L of the balls $B(T_2 r_j, x_j)$.

We fix once and for all a function $\theta \in C_0^\infty(B(2,0))$ such that $0 \le \theta \le 1$ and $\theta = 1$ on $B(1,0)$, and we set $\theta_i(x) = \theta((x_i^{-1}x)/r_i)$. Then $\operatorname{supp} \theta_i \subset B(2r_i, x_i)$ and $1 \le \Sigma_i \theta_i(x) \le L$ for $x \in \Omega$. Thus if we set

$$\zeta_i(x) = \theta_i(x)/\Sigma_j\theta_j(x) \qquad (x \in \Omega)$$

$$= 0 \qquad (x \in \Omega^c),$$

then $\zeta_i \in C_0^\infty(B(2r_i, x_i))$, $0 \le \zeta_i \le 1$, $\zeta_i = 1$ on $B(r_i/4\gamma, x_i)$, and $\Sigma_i\zeta_i = \chi_\Omega$.

Now let us fix $a \in \Delta$. For each i, we equip the space $\mathcal{P}_a$ with the Hilbert space norm

$$\|P\|^2 = \left(\int \zeta_i\right)^{-1} \int |P(x)|^2 \zeta_i(x)dx. \tag{3.5}$$

Our distribution f defines a linear functional on this Hilbert space by

$$P \longrightarrow \left(\int \zeta_i\right)^{-1} \langle f, P\zeta_i \rangle,$$

which is therefore represented by a unique $P_i \in \mathcal{P}_a$. That is, P_i is specified by the condition

$$\int f(x)Q(x)\zeta_i(x)dx = \int P_i(x)Q(x)\zeta_i(x)dx \qquad \text{for all } Q \in \mathcal{P}_a.$$

For each i, we define

$$b_i = (f - P_i)\zeta_i.$$

It will turn out, in the cases of interest, that the series Σb_i converges in $\mathcal{S}'$. In this case, we set

$$g = f - \Sigma_i b_i,$$

and we call the representation

$$f = g + \Sigma_i b_i$$

a *Calderón-Zygmund decomposition of* f *of degree* a *and height* α *associated to* $M_{(N)}f$.

We proceed to explore the properties of the functions b_i. Henceforth we shall use the notation

$$B_i = B(r_i, x_i), \quad \tilde{B}_i = B(T_1 r_i, x_i).$$

(3.6) LEMMA. *If* $\tilde{B}_i \cap \tilde{B}_j \neq \emptyset$ *then* $r_i \leq (1+6\gamma^2) r_j$.

Proof: If $\tilde{B}_i \cap \tilde{B}_j \neq \emptyset$ then $|x_i^{-1} x_j| < \gamma T_1(r_i + r_j)$, so by (3.3),

$$\begin{aligned} 2\gamma^2 T_1 r_i = T_2 r_i &\leq \operatorname{dist}(x_i, \Omega^c) \leq \gamma(|x_i^{-1}x_j| + \operatorname{dist}(x_j, \Omega^c)) \\ &\leq \gamma(T_1\gamma(r_i + r_j) + T_3 r_j) \\ &= T_1\gamma^2 r_i + T_1\gamma^2(1+6\gamma^2) r_j. \end{aligned}$$

The lemma follows on solving for r_i. #

(3.7) LEMMA. *There is a constant* A_1, *depending only on* N, *such that*

$$\sup_{|I| \leq N, x \in G} r_i^{d(I)} |Y^I \zeta_i(y)| \leq A_1.$$

<u>Proof</u>: Given i, let $J = \{j : \tilde{B}_j \cap \tilde{B}_i \neq \emptyset\}$. Then $\text{card}(J) \leq L$ by (3.4), and $\zeta_i = \theta_i / \Sigma_{j \in J}\, \theta_j$. Moreover, $\theta_j(x) = \theta((x_j^{-1}x)/r_j)$ and hence for $j \in J$ and $|I| \leq N$, by Lemma 3.6 we have

$$\sup_y |Y^I \theta_j(y)| = r_j^{-d(I)} \sup_y |Y^I \theta(y)| \leq C r_j^{-d(I)} \leq C' r_i^{-d(I)}.$$

The lemma now follows easily from the product rule for derivatives. #

(3.8) LEMMA. <u>There is a constant</u> A_2, <u>independent of</u> f, i, <u>and</u> α, <u>such that</u>

$$\sup_y |P_i(y)\zeta_i(y)| \leq A_2 \alpha.$$

<u>Proof</u>: Let $\pi_1, \ldots, \pi_m$ ($m = \dim P_a$) be an orthonormal basis for P_a with respect to the norm (3.5). Then by the properties of ζ_i,

$$1 = \left(\int \zeta_i\right)^{-1} \int |\pi_j(y)|^2 \zeta_i(y)\,dy \geq |B(2r_i, x_i)|^{-1} \int_{B(r_i/4\gamma, x_i)} |\pi_j(y)|^2 dy$$

$$= 2^{-Q} \int_{B(1/4\gamma, 0)} |\tilde{\pi}_j(y)|^2 dy$$

where $\tilde{\pi}_j(z) = \pi_j(x(r_i z))$. Since $\dim P_a < \infty$ (so that all norms on P_a are equivalent), there exists $C_1 > 0$ such that for all $P \in P_a$,

$$\sup_{|I| \leq N, |z| \leq 2} |Y^I P(z)| \leq C_1 \left(\int_{B(1/4\gamma, 0)} |P(z)|^2 dz\right)^{1/2}.$$

Applying this to $\tilde{\pi}_j$, we obtain

(3.9) $$\sup_{|I| \leq N, y \in B(2r_i, x_i)} r_i^{d(I)} |Y^I \pi_j(y)| \leq 2^{Q/2} C_1.$$

Combining this with Lemma 3.7, we have

$$(3.10) \qquad \sup_{|I|\le N, y\in G} r_i^{d(I)} |Y^I(\pi_j\zeta_i)(y)| \le C_2.$$

Now choose $z \in B(T_3 r_i, x_i) \cap \Omega^c$ and set

$$\Phi_j(y) = \left(\int \zeta_i\right)^{-1} r_i^Q \pi_j(z(r_i y^{-1}))\zeta_i(z(r_i y^{-1})).$$

Then $\operatorname{supp} \Phi_j \subset B(T_3+2\gamma, 0)$ and the estimate (3.10) implies that

$$\sup_{|I|\le N, y\in G} |Y^I \Phi_j(y)| \le C_3.$$

Therefore $\|\Phi_j\|_{(N)} \le C_4$, so since $z \notin \Omega$,

$$(3.11) \qquad \left|\left(\int \zeta_i\right)^{-1} \int f(x)\pi_j(x)\zeta_i(x)dx\right| = |f*\Phi_{r_i}(z)| \le C_4\alpha.$$

But

$$P_i = \Sigma_1^m\left[\left(\int \zeta_i\right)^{-1} \int f(x)\pi_j(x)\zeta_i(x)dx\right]\bar{\pi}_j,$$

so the lemma follows from (3.9) (with $I = 0$) and (3.11). #

(3.12) LEMMA. <u>There is a constant</u> A_3, <u>independent of</u> f, i, <u>and</u> α, <u>such that</u>

$$Mb_i(x) \le A_3 Mf(x) \quad \underline{\text{for}} \quad x \in \tilde{B}_i.$$

<u>Proof</u>: Suppose $\phi \in S$, $\|\phi\|_{(N)} \le 1$, $t > 0$, and $x \in \tilde{B}_i$. If $t \le r_i$, we write $b_i*\phi_t(x) = f*\Phi_t(x) - (P_i\zeta_i)*\phi_t(x)$ where $\Phi(z) = \phi(z)\zeta_i(x(tz^{-1}))$. It follows from Lemma 3.7 that $\|\Phi\|_{(N)} \le C_1$ (independent of i and α),

and hence by Lemma 3.8, since $x \in \Omega$,

$$|b_i * \phi_t(x)| \leq C_1 Mf(x) + A_2 \alpha \|\phi\|_1$$

$$\leq C_2(Mf(x) + \alpha\|\phi\|_{(N)})$$

$$\leq (C_2+1)Mf(x).$$

On the other hand, if $t > r_i$ we write

$$b_i * \phi_t(x) = (r_i/t)^Q f * \Phi_{r_i}(x) - (P_i\zeta_i) * \phi_t(x)$$

where $\Phi(z) = \phi((r_i/t)z)\zeta(x(r_i z^{-1}))$. If $\Phi(z) \neq 0$ we must have $|x_i^{-1}x(r_i z^{-1})| \leq 2r_i$, which implies that

$$|z| \leq \gamma(T_1 + r_i^{-1}|x_i^{-1}x|) \leq \gamma(T_1+2),$$

so supp $\Phi \subset B(\gamma(T_1+2),0)$. In view of this fact and the inequality $r_i/t \leq 1$, it follows as above that $\|\Phi\|_{(N)} \leq C$, and hence

$$|b_i * \phi_t(x)| \leq C(r_i/t)^Q Mf(x) + A_2 \alpha \|\phi\|_1$$

$$\leq C' Mf(x).$$

Since $Mb_i \leq CM^0 b_i$, the lemma follows immediately. #

(3.13) LEMMA. *There is a constant* A_4, *independent of* f, i, *and* α, *such that*

$$Mb_i(x) \leq A_4 \alpha (r_i/|x_i^{-1}x|)^{Q+b} \qquad \textit{for } x \notin \tilde{B}_i,$$

where $b = \min\{b' \in \Delta : b' > a\}$ *if* $a < N$, *and* $b = N$ *if* $a \geq N$.

<u>Proof</u>: Suppose $\phi \in \mathcal{S}$, $\|\phi\|_{(N)} \leq 1$, and $t > 0$. Pick $w \in B(T_3 r_i, x_i) \cap \Omega^c$.

<u>Case I</u>: $t \leq r_i$. We write $b_i * \phi_t(x) = f * \Phi_t(w) - (P_i \zeta_i) * \phi_t(x)$ where

$$\Phi(z) = \phi\left(z\left(\frac{w^{-1}x}{t}\right)\right)\zeta_i(w(tz^{-1})).$$

If $z \in \operatorname{supp} \Phi$ then $|x_i^{-1} w(tz^{-1})| \leq 2r_i$, whence

$$|z| \leq (\gamma/t)(2r_i + T_3 r_i) \leq Cr_i/t,$$

and since $|x_i^{-1}x| \geq T_1 r_i \geq 4\gamma r_i$,

$$\left|z\left(\frac{w^{-1}x}{t}\right)\right| = \left|\frac{x_i^{-1}x}{t}\right| - \left(\left|\frac{x_i^{-1}x}{t}\right| - \left|z\left(\frac{w^{-1}x}{t}\right)\right|\right)$$

$$\geq \left|\frac{x_i^{-1}x}{t}\right| - \gamma\left|z\left(\frac{w^{-1}x_i}{t}\right)\right|$$

$$\geq \left|\frac{x_i^{-1}x}{t}\right| - \frac{2\gamma r_i}{t}$$

$$\geq \frac{1}{2}\left|\frac{x_i^{-1}x}{t}\right|.$$

Therefore, if $|I| \leq N$, by Lemma 3.7 we have

$$(1+|z|)^{(N+1)(Q+1)} |Y^I \Phi(z)|$$

$$\leq C_2 \sum_{|J| \leq |I|} \left(\frac{r_i}{t}\right)^{(N+1)(Q+1)} \left(\frac{|x_i^{-1}x|}{t}\right)^{-(N+1)(Q+1)} \left(\frac{t}{r_i}\right)^{d(I)}$$

$$\leq C_3 (r_i/|x_i^{-1}x|)^{(N+1)(Q+1)}$$

since $t/r_i \leq 1$. Also, if $y \in \operatorname{supp} b_i$ and $x \notin \tilde{B}_i$ we have $|y^{-1}x| \geq C_4|x_i^{-1}x|$, so by Lemma 3.8,

$$|b_i * \phi_t(x)|$$

$$\leq C_3(r_i/|x_i^{-1}x|)^{(N+1)(Q+1)}Mf(w)+A_2\alpha t^{-Q}\int_{B(2r_i,x_i)} |(y^{-1}x)/t|^{-(N+1)(Q+1)}d$$

$$\leq C_3\alpha(r_i/|x_i^{-1}x|)^{(N+1)(Q+1)} + C_5\alpha t^{(N+1)(Q+1)-Q}\, r_i^Q|x_i^{-1}x|^{-(N+1)(Q+1)}$$

$$\leq C_6\alpha(r_i/|x_i^{-1}x|)^{(N+1)(Q+1)}$$

$$\leq C_6\alpha(r_i/|x_i^{-1}x|)^{Q+b}$$

since $t \leq r_i$, $b \leq N$, and $r_i < |x_i^{-1}x|$.

Case II: $t > r_i$. Let $a' = a$ if $a < N$ and $a' = \min\{a'' \in \Delta : a'' < N\}$ if $a \geq N$; thus $a' = \max\{a'' \in \Delta : a'' < b\}$ and $N \geq [a']+1$. For any $z \in G$ we write $\phi(yz) = P_z(y)+R_z(y)$ where P_z is the right Taylor polynomial of ϕ at z of homogeneous degree a'. By Theorem 1.50, R_z satisfies

$$(3.14) \qquad |Y^I R_z(y)| \leq C_7|y|^{b-d(I)}|z|^{-Q-b} \qquad (d(I) < b, \quad |z| \geq 2\gamma\beta^N|y|).$$

Also, if $d(I) \geq b$ then $Y^I R_z(y) = Y^I\phi(yz)$, so

$$(3.15) \qquad |Y^I R_z(y)| \leq (1+|yz|)^{-(N+1)(Q+1)}$$

$$\leq C_8|z|^{-Q-d(I)} \qquad (|I| \leq N, \quad d(I) \geq b, \quad |z| \geq 2\gamma|y|).$$

Now, by the construction of b_i we have

$$b_i*\phi_t(x) = t^{-Q}\int b_i(y)\, R_{(x_i^{-1}x)/t}((yx_i^{-1})/t)dy$$

$$= (r_i/t)^Q f*\Phi_{r_i}(w) + t^{-Q}\int P_i(y)\zeta_i(y)R_{(x_i^{-1}x)/t}((yx_i^{-1})/t)dy,$$

where

$$\Phi(z) = R_{(x_i^{-1}x)/t}\left(\frac{(r_iz)w^{-1}x_i}{t}\right)\zeta_i(w(r_iz^{-1})).$$

If $z \in \operatorname{supp}\Phi$ we have, as before, $|z| \leq C_1r_i/t$, which now implies that $|z| \leq C_1$, and

(3.16) $$2\gamma\beta^N\left|\frac{(r_iz)w^{-1}x_i}{t}\right| < \frac{T_1}{2t}|x_iw(r_iz^{-1})| \leq \frac{T_1r_i}{t} \leq \left|\frac{x_i^{-1}x}{t}\right|.$$

By Lemma 3.7, the derivatives of $\zeta_i(w(r_iz^{-1}))$ with respect to z are bounded by a constant. Therefore, by the estimates (3.14), (3.15), and (3.16), for $|J| \leq N$ we have

$$|Y^J\Phi(z)| \leq C_9\sum_{|I|\leq|J|,d(I)<b}\left(\frac{r_i}{t}\right)^{d(I)}\left|\frac{(r_iz)w^{-1}x_i}{t}\right|^{b-d(I)}\left|\frac{x_i^{-1}x}{t}\right|^{-Q-b}$$

$$+ C_{10}\sum_{|I|\leq|J|,d(I)\geq b}\left(\frac{r_i}{t}\right)^{d(I)}\left|\frac{x_i^{-1}x}{t}\right|^{-Q-d(I)}$$

$$\leq C_{11}\left(\frac{r_i}{t}\right)^{b}\left|\frac{x_i^{-1}x}{t}\right|^{-Q-b}$$

$$+ C_{10}\sum_{|I|\leq|J|,d(I)\geq b}\left(\frac{t}{r_i}\right)^{Q}\left(\frac{r_i}{|x_i^{-1}x|}\right)^{Q+d(I)}$$

$$\leq C_{12}(t/r_i)^Q(r_i/|x_i^{-1}x|)^{Q+b},$$

since $|x_i^{-1}x| > r_i$. Therefore, since $\operatorname{supp} \Phi \subset B(C_1,0)$, we have

$$\|\Phi\|_{(N)} \leq C_{13}(t/r_i)^Q(r_i/|x_i^{-1}x|)^{Q+b}$$

and hence

$$|(r_i/t)^Q f*\Phi_{r_i}(w)| \leq C_{13}\alpha(r_i/|x_i^{-1}x|)^{Q+b}.$$

On the other hand, by (3.14) and Lemma 3.8,

$$|t^{-Q}\int P_i(y)\zeta_i(y)R_{(x_i^{-1}x)/t}((yx_i^{-1})/t)dy|$$

$$\leq C_{14}\alpha t^{-Q}\int_{B(2r_i,x_i)}\left|\frac{yx_i^{-1}}{t}\right|^b\left|\frac{x_i^{-1}x}{t}\right|^{-Q-b}dy$$

$$\leq C_{15}\alpha(r_i/|x_i^{-1}x|)^{Q+b}.$$

Combining these last two estimates, we are done. #

(3.17) THEOREM. <u>Suppose</u> $0 < p \leq 1$, $N = N_p$, a <u>is p-admissible and</u> $f \in H^p$. <u>There is a constant</u> A_5, <u>independent of</u> f, i, <u>and</u> α, <u>such that</u>

$$\int Mb_i(x)^p dx \leq A_5\int_{\tilde{B}_i} Mf(x)^p dx.$$

<u>Moreover, the series</u> Σb_i <u>converges in</u> H^p, <u>and if</u> L <u>is as in</u> (3.4),

$$\int M(\Sigma b_i)(x)^p dx \leq LA_5\int_{\Omega} Mf(x)^p dx.$$

<u>Proof</u>: By Lemmas 3.12 and 3.13, we have

$$\int Mb_i(x)^p dx \leq A_3^p\int_{\tilde{B}_i} Mf(x)^p dx + A_4^p\alpha^p\int_{\tilde{B}_i^c}(r_i/|x_i^{-1}x|)^{p(Q+b)}dx$$

where $b > Q(p^{-1}-1)$. Hence $p(Q+b) > Q$, so

$$\int_{\tilde{B}_i^c} (r_i/|x_i^{-1}x|)^{p(Q+b)} \leq C_1 r_i^Q \leq C_1|\tilde{B}_i|.$$

Hence, since $\tilde{B}_i \subset \Omega$,

$$\int Mb_i(x)^p dx \leq C_2[\int_{\tilde{B}_i} Mf(x)^p dx + \alpha^p|\tilde{B}_i|]$$

$$\leq 2C_2 \int_{\tilde{B}_i} Mf(x)^p dx.$$

This proves the first assertion, and since H^p is complete, the second follows from the estimate

$$\Sigma_i \int Mb_i(x)^p dx \leq 2C_2\Sigma_i \int_{\tilde{B}_i} Mf(x)^p dx \leq 2LC \int_\Omega Mf(x)^p dx. \quad \#$$

<u>Remark</u>: If we replace the assumptions $N = N_p$, $f \in H^p$ by $N \geq N_p$, $Mf(=M_{(N)}f) \in L^p$, we obtain the same conclusions, except that the series Σb_i converges in the topology defined by the maximal operator $M_{(N)}$.

(3.18) THEOREM. <u>Suppose</u> $N \geq 0$, $a \in \Delta$, <u>and</u> $f \in L^1$. <u>Then the series</u> Σb_i <u>converges in</u> L^1, <u>and there is a constant</u> A_6, <u>independent of</u> f, i, <u>and</u> α, <u>such that</u>

$$\int \Sigma |b_i(x)| dx \leq A_6 \int |f(x)| dx.$$

<u>Proof</u>: By Lemma 3.8,

$$\int |b_i| = \int |(f-P_i)\zeta_i| \leq \int_{\tilde{B}_i} |f| + \int |P_i\zeta_i|$$

$$\leq \int_{\tilde{B}_i} |f| + A_2\alpha|\tilde{B}_i|.$$

Hence by (3.4) and the maximal theorem (2.4),

$$\Sigma \int |b_i| \leq L \int_{\Omega} |f| + LA_2\alpha|\Omega|$$

$$\leq C \int |f|. \quad \#$$

We thus have Calderón-Zygmund decompositions for $f \in S'$ such that $M_{(N)}f \in L^p$ $(N \geq N_p,\ 0 < p \leq 1)$ and for $f \in L^1$. We now investigate the "good part" $g = f - \Sigma b_i$.

(3.19) LEMMA. <u>Suppose</u> Σb_i <u>converges in</u> S'. <u>There is a constant</u> A_7, <u>independent of</u> f <u>and</u> α, <u>such that for all</u> $x \in G$,

$$Mg(x) \leq A_7\alpha\Sigma_i\left(\frac{r_i}{|x_i^{-1}x|+r_i}\right)^{Q+b} + Mf(x)\chi_{\Omega^c}(x),$$

<u>where</u> b <u>is as in Lemma</u> 3.13.

<u>Proof</u>: If $x \notin \Omega$, by Lemma 3.13 we have

$$Mg(x) \leq Mf(x) + \Sigma_i Mb_i(x)$$

$$\leq Mf(x) + \Sigma_i A_4\alpha(r_i/|x_i^{-1}x|)^{Q+b}$$

$$\leq Mf(x) + 2^{Q+b}A_4\alpha\Sigma_i\left(\frac{r_i}{|x_i^{-1}x|+r_i}\right)^{Q+b}$$

since $|x_i^{-1}x| > r_i$. On the other hand, if $x \in \Omega$ let us choose k such that $x \in B_k$, and let $J = \{i : \tilde{B}_i \cap \tilde{B}_k \neq \emptyset\}$. Then $\text{card}(J) \leq L$, and as above we have

$$\Sigma_{i\notin J} Mb_i(x) \leq A_4 \alpha \Sigma_{i\notin J}(r_i/|x_i^{-1}x|)^{Q+b}$$

$$\leq 2^{Q+b} A_4 \alpha \Sigma_{i\notin J}\left(\frac{r_i}{|x_i^{-1}x|+r_i}\right)^{Q+b}.$$

Hence it suffices to estimate the maximal function of $g+\Sigma_{i\notin J} b_i = f-\Sigma_{i\in J} b_i$. As in the proof of Lemma 3.13, we fix $w \in B(T_3 r_k, x_k) \cap \Omega^c$.

Suppose $\phi \in S$, $\|\phi\|_{(N)} \leq 1$, and $t > 0$. If $t \leq r_k$, we write

$$(f-\Sigma_{i\in J} b_i)*\phi_t(x) = (f\eta)*\phi_t(x) + (\Sigma_{i\in J} P_i \zeta_i)*\phi_t(x)$$

$$= f*\Phi_t(w) + (\Sigma_{i\in J} P_i \zeta_i)*\phi_t(x)$$

where

$$\eta = 1-\Sigma_{i\in J}\zeta_i, \qquad \Phi(z) = \phi(z(\tfrac{w^{-1}x}{t}))\eta(w(tz^{-1})).$$

We observe that $\eta = 0$ on $\tilde{B}_k$, so if $z \in \operatorname{supp} \Phi$ and $y = z((w^{-1}x)/t)$,

$$|y| = t^{-1}|(tz)w^{-1}x_k x_k^{-1}x| \geq t^{-1}(T_1 r_k - \gamma r_k) \geq C_1 r_k/t$$

where $C_1 > 0$, and hence

$$|z| \leq \gamma(|y| + |w^{-1}x|/t) \leq \gamma(|y| + T_3 r_k/t) \leq C_2|y|.$$

Therefore, by Lemma 3.7, for $|I| \leq N$ we have

$$(1+|z|)^{(Q+1)(N+1)} \ |Y^I\Phi(z)|$$

$$\leq A_1(1+|z|)^{(Q+1)(N+1)}(1+|y|)^{-(Q+1)(N+1)} \ \Sigma_{|J|\leq|I|} \ (t/r_k)^{d(J)}$$

$$\leq C_3,$$

so $\|\Phi\|_{(N)} \leq C_3$ and

$$|f*\Phi_t(w)| \leq C_3\alpha \leq 2^{Q+b}C_3\alpha(\frac{r_k}{|x_k^{-1}x|+r_k})^{Q+b}.$$

Also, by (3.4) and Lemma 3.8,

$$|(\Sigma_{i\in J}P_i\zeta_i)*\phi_t(x)| \leq LA_2\alpha \int|\phi| \leq C_4\alpha \leq 2^{Q+b}C_4\alpha(\frac{r_k}{|x_k^{-1}x|+r_k})^{Q+b}.$$

This completes the estimate for $t \leq r_k$.

If $t > r_k$, let $\Phi(z) = \phi(z((w^{-1}x)/t))$. Since $|w^{-1}x|/t \leq T_3r_k/t \leq T_3$, by Proposition 1.46 we have $\|\Phi\|_{(N)} \leq C_5$, so

$$|f*\phi_t(x)| = |f*\Phi_t(w)| \leq C_5\alpha \leq 2^{Q+b}C_5\alpha(\frac{r_k}{|x_k^{-1}x|+r_k})^{Q+b}.$$

Also, by Lemma 3.13,

$$\Sigma_{i\in J}|b_i*\phi_t(x)| = \Sigma_{i\in J}|b_i*\Phi_t(w)| \leq C_5\Sigma_{i\in J}Mb_i(w)$$

$$\leq A_4C_5\alpha\Sigma_{i\in J}(r_i/|x_i^{-1}w|)^{Q+b}.$$

Now, by Lemma 3.6, for $i \in J$ we have

$$|x_i^{-1}x| \le \gamma(|x_i^{-1}x_k| + |x_k^{-1}x|) \le \gamma[T_1\gamma(r_i+r_k) + r_k]$$

$$\le \gamma[T_1\gamma r_i + (1+T_1\gamma)(1+6\gamma^2)r_i] = C_6 r_i.$$

On the other hand, $|x_i^{-1}w| > T_1 r_i$ since $w \notin \Omega$. Hence

$$|x_i^{-1}x| + r_i \le (C_6+1)r_i \le T_1^{-1}(C_6+1)|x_i^{-1}w|,$$

so that

$$\Sigma_{i\in J}|b_i * \phi_t(x)| \le C_7\alpha\Sigma_{i\in J}\left(\frac{r_i}{|x_i^{-1}x|+r_i}\right)^{Q+b},$$

and we are done. #

(3.20) THEOREM. (i) <u>Suppose</u> $0 < p \le 1$, $N \ge N_p$, a <u>is</u> p-<u>admissible</u>, <u>and</u> $Mf \in L^p$. <u>Then</u> $Mg \in L^1$, <u>and there is a constant</u> A_8, <u>independent of</u> f <u>and</u> α, <u>such that</u>

$$\int Mg(x)dx \le A_8\alpha^{1-p}\int Mf(x)^p dx.$$

(ii) <u>Suppose</u> $N \ge 0$, $a \in \Delta$, <u>and</u> $f \in L^1$. <u>Then</u> $g \in L^\infty$, <u>and there is a constant</u> A_9, <u>independent of</u> f <u>and</u> α, <u>such that</u> $\|g\|_\infty \le A_9\alpha$.

<u>Proof</u>: (i) By Lemma 3.19,

$$\int Mg(x)dx \le A_7\alpha\Sigma_i\int\left(\frac{r_i}{|x_i^{-1}x|+r_i}\right)^{Q+b}dx + \int_{\Omega^c} Mf(x)dx. \tag{3.21}$$

Let

$$C = \int_0^\infty s^{Q-1}(s+1)^{-Q-b}ds.$$

Then the first term on the right of (3.21) is bounded by

$$A_7 C\alpha \Sigma_i r_i^Q = A_7 C\alpha \Sigma_i |B_i| \le LA_7 C\alpha|\Omega| = C'\alpha|\Omega|.$$

Hence

$$\begin{aligned}\int Mg(x)dx &\le C'\alpha|\Omega| + \int_{\Omega^c} Mf(x)dx \\ &\le C'\alpha\cdot\alpha^{-p}\int_{\Omega} Mf(x)^p dx + \alpha^{1-p}\int_{\Omega^c} Mf(x)^p dx \\ &\le C''\alpha^{1-p}\int Mf(x)^p dx.\end{aligned}$$

(ii) If $f \in L^1$ then g and the b_i's are functions, and

$$g = f - \Sigma b_i = f(1-\Sigma\zeta_i) + \Sigma P_i\zeta_i = f\chi_{\Omega^c} + \Sigma P_i \zeta_i.$$

Thus by Lemma 3.8, for $x \in \Omega$ we have $|g(x)| \le LA_2\alpha$, while by Theorem 2.6, for almost every $x \in \Omega^c$ we have $|g(x)| = |f(x)| \le CMf(x) \le C\alpha$. #

This completes our discussion of the Calderón-Zygmund decomposition. As an immediate corollary, we obtain the following important result.

(3.22) COROLLARY. _If_ $0 < p < 1$, $H^p \cap L^1$ _is dense in_ H^p.

Proof: If $f \in H^p$ and $\alpha > 0$, let $f = g^\alpha + \Sigma b_i^\alpha$ be a Calderón-Zygmund decomposition of f of degree a and height α associated to $Mf = M_{(N_p)}f$, were a is p-admissible. By Theorem 3.17,

$$\rho^p(\Sigma_i b_i^\alpha) \le C\int_{\{x:Mf(x)>\alpha\}} Mf(x)^p dx,$$

so $\rho^p(\Sigma_i b_i^\alpha) \to 0$ as $\alpha \to \infty$. Hence $g^\alpha \to f$ in H^p as $\alpha \to \infty$, but $g^\alpha \in L^1$ by Theorems 3.20 and 2.7. #

Remark: The same proof shows that if $N \geq N_p$ and $M_{(N)}f \in L^p$ then the "good parts" g^α of the Calderón-Zygmund decompositions of f associated to $M_{(N)}f$ are in L^1 and

$$\int [M_{(N)}(f-g^\alpha)(x)]^p dx \to 0 \quad \text{as} \quad \alpha \to \infty.$$

B. The Atomic Decomposition

We now aim to prove that Hardy spaces coincide with atomic Hardy spaces. Suppose $0 < p \leq 1$, $N \geq N_p$, a is p-admissible, and f is a distribution such that $M_{(N)}f \in L^p$. For each $k \in \mathbb{Z}$, let $f = g^k + \Sigma_i b_i^k$ be a Calderón-Zygmund decomposition of f of degree a and height 2^k associated to $M_{(N)}f$. We shall label all the ingredients in this construction as in Section A, but with superscript k's: for example,

$$\Omega^k = \{x : M_{(N)}f(x) > 2^k\}, \quad b_i^k = (f-P_i^k)\zeta_i^k, \quad B_i^k = B(r_i^k, x_i^k).$$

We now need two more definitions. First, by analogy with the definition of P_i^k, we define P_{ij}^{k+1} to be the orthogonal projection of $(f-P_j^{k+1})\zeta_i^k$ on P_a with respect to the norm

$$\|P\|^2 = \left(\int \zeta_j^{k+1}\right)^{-1} \int |P(x)|^2 \zeta_j^{k+1}(x)dx. \tag{3.23}$$

That is, P_{ij}^{k+1} is the unique element of P_a such that for all $Q \in P_a$,

$$\int (f-P_j^{k+1})\zeta_i^k Q \zeta_j^{k+1} = \int P_{ij}^{k+1} Q \zeta_j^{k+1}.$$

Second, we define

$$\hat{B}_i^k = B(2r_i^k, x_i^k).$$

(3.24) LEMMA. (a) *If* $\hat{B}_j^{k+1} \cap \hat{B}_i^k \neq \emptyset$ *then* $r_j^{k+1} < 4\gamma r_i^k$ *and* $\hat{B}_j^{k+1} \subset B(T_2 r_i^k, x_i^k)$. (b) *For each* j *there are at most* L *values of* i *for which* $\hat{B}_j^{k+1} \cap \hat{B}_i^k \neq \emptyset$, *where* L *is as in* (3.4).

Proof: If $\hat{B}_j^{k+1} \cap \hat{B}_i^k \neq \emptyset$ we have

(3.25) $$|(x_i^k)^{-1} x_j^{k+1}| < 2\gamma(r_i^k + r_j^{k+1}),$$

and since $\Omega^{k+1} \subset \Omega^k$,

$$\operatorname{dist}(x_j^{k+1}, (\Omega^k)^c) \geq \operatorname{dist}(x_j^{k+1}, (\Omega^{k+1})^c) \geq T_2 r_j^{k+1}.$$

Thus

$$\begin{aligned} 18\gamma^3\beta^N r_j^{k+1} = T_2 r_j^{k+1} &\leq \operatorname{dist}(x_j^{k+1}, (\Omega^k)^c) \\ &\leq \gamma[|(x_i^k)^{-1}(x_j^{k+1})| + \operatorname{dist}(x_i^k, (\Omega^k)^c)] \\ &\leq \gamma[2\gamma(r_j^{k+1} + r_i^k) + T_3 r_i^k] \\ &= 2\gamma^2 r_j^{k+1} + 2\gamma^2(1 + 27\gamma^2\beta^N) r_i^k, \end{aligned}$$

so that

$$r_j^{k+1} \leq (27\gamma^2\beta^N + 1)(9\gamma\beta^N - 1)^{-1} r_i^k \leq (28/8)\gamma r_i^k < 4\gamma r_i^k.$$

From this and (3.25) it follows that

$$|(x_i^k)^{-1} x_j^{k+1}| \leq 2\gamma(1+4\gamma) r_i^k,$$

so that if $y \in \hat{B}_j^{k+1}$,

$$|(x_i^k)^{-1}y| < \gamma[2\gamma(1+4\gamma)r_i^k + 2r_j^{k+1}] < (2\gamma^2+8\gamma^3+8\gamma)r_i^k \le T_2 r_i^k .$$

This proves (a), and (b) follows from (a) and (3.4). #

(3.26) LEMMA. <u>There is a constant</u> A_{10}, <u>independent of</u> i, j, <u>and</u> k, such that

$$\sup_y |P_{ij}^{k+1}(y)\zeta_j^{k+1}(y)| \le A_{10}2^{k+1}.$$

<u>Proof</u>: The argument is essentially the same as the proof of Lemma 3.8, and we indicate only the necessary modifications. If $\pi_1,\ldots,\pi_m$ is an orthonormal basis for $\mathcal{P}_a$ with respect to the norm (3.23), it suffices to show that for some $C > 0$ independent of i, j, k, and ℓ,

$$|(\int \zeta_j^{k+1})^{-1} \int (f-P_j^{k+1})\zeta_i^k \pi_\ell \zeta_j^{k+1}| \le C2^{k+1}.$$

But by Lemma 3.8 and its proof (with P_i,ζ_i replaced by P_j^{k+1}, ζ_j^{k+1}),

$$|P_j^{k+1}(y)| \le C2^{k+1}, \quad |\pi_\ell(y)| \le C \quad \text{for } y \in \hat{B}_j^{k+1}.$$

Hence

$$|(\int \zeta_j^{k+1})^{-1} \int P_j^{k+1}\pi_\ell \zeta_i^k \zeta_j^{k+1}| \le C2^{k+1},$$

so we need to show that

$$|(\int \zeta_j^{k+1})^{-1} \int f\pi_\ell \zeta_i^k \zeta_j^{k+1}| \le C2^{k+1}.$$

Now

$$\left(\int \zeta_j^{k+1}\right)^{-1} \int f\pi_\ell \zeta_i^k \zeta_j^{k+1} = f*\Phi_{r_j^{k+1}}(w)$$

where $w \in B(T_3 r_j^{k+1}, x_j^{k+1}) \cap (\Omega^{k+1})^c$ and

$$\Phi(z) = (r_j^{k+1})^Q \left(\int \zeta_j^{k+1}\right)^{-1} [\pi_\ell \zeta_i^k \zeta_j^{k+1}](w(r_j^{k+1} z^{-1})),$$

so it suffices to show that $\|\Phi\|_{(N)} \leq C$. However, we need only consider those values of i and j such that $\hat{B}_j^{k+1} \cap \hat{B}_i^k \neq \emptyset$, since otherwise $\zeta_i^k \zeta_j^{k+1}$ vanishes identically, and for these values of i and j, Lemmas 3.7 and 3.24 yield the desired result. #

(3.27) LEMMA. <u>For every</u> $k \in \mathbb{Z}$, $\Sigma_i(\Sigma_j P_{ij}^{k+1} \zeta_j^{k+1}) = 0$, <u>where the series converges pointwise and in</u> S'.

<u>Proof</u>: For each x there are at most L values of j for which $\zeta_j^{k+1}(x) \neq 0$. Moreover, for each j, P_j^{k+1} is zero unless $\hat{B}_i^k \cap \hat{B}_j^{k+1} \neq \emptyset$, and by Lemma 3.24 there are at most L values of i for which this happens. Thus for each x the series $\Sigma_i \Sigma_j P_{ij}^{k+1}(x) \zeta_j^{k+1}(x)$ is actually a finite sum, and by Lemma 3.26,

$$\Sigma_i \Sigma_j |P_{ij}^{k+1}(x) \zeta_j^{k+1}(x)| \leq CL^2 2^{k+1}.$$

From this it follows easily that the series $\Sigma_i(\Sigma_j P_{ij}^{k+1} \zeta_j^{k+1})$ and $\Sigma_j(\Sigma_i P_{ij}^{k+1}) \zeta_j^{k+1}$ converge in S' and are equal, so it suffices to see that $\Sigma_i P_{ij}^{k+1} = 0$ for each j. Given j, let $I = \{i : \hat{B}_i^k \cap \hat{B}_j^{k+1} \neq \emptyset\}$. Then $\Sigma_i P_{ij}^{k+1}$ is the projection of $(f - P_j^{k+1}) \Sigma_{i \in I} \zeta_i^k$ on $\mathcal{P}_a$ with respect to the norm (3.23). But $\Sigma_{i \in I} \zeta_i^k = 1$ on $\hat{B}_j^{k+1}$, so $\Sigma_i P_{ij}^{k+1}$ is the projection of $f - P_j^{k+1}$, which is zero by definition of P_j^{k+1}. #

(3.28) THEOREM. <u>If</u> $0 < p \le 1$ <u>and</u> a <u>is</u> p-<u>admissible then</u> $H^p \subset H^p_{\infty,a}$, <u>and the inclusion is continuous</u>.

<u>Proof</u>: Suppose first that $f \in H^p \cap L^1$. Let $f = g^k + \Sigma_i b_i^k$ be the Calderón-Zygmund decompositions of f as in the preceding lemmas, where $N = N_p$. By Theorem 3.17, $g^k \to f$ in H^p as $k \to +\infty$, while by Theorem 3.20 (since $f \in L^1$), $g^k \to 0$ uniformly as $k \to -\infty$. Therefore

$$f = \sum_{-\infty}^{\infty}(g^{k+1}-g^k) \qquad \text{(convergence in } S'\text{)}.$$

Now, using Lemma 3.27 together with the equation

$$\Sigma_i \zeta_i^k b_j^{k+1} = \chi_{\Omega^k} b_j^{k+1} = b_j^{k+1}$$

we have

$$g^{k+1}-g^k = (f-\Sigma_j b_j^{k+1}) - (f-\Sigma_i b_i^k)$$

$$= \Sigma_i b_i^k - \Sigma_j b_j^{k+1} + \Sigma_i \Sigma_j P_{ij}^{k+1} \zeta_j^{k+1}$$

$$= \Sigma_i [b_i^k - \Sigma_j (\zeta_i^k b_j^{k+1} - P_{ij}^{k+1} \zeta_j^{k+1})]$$

$$= \Sigma_i h_i^k$$

where all the series converge in S' and

(3.29) $$h_i^k = (f-P_i^k)\zeta_i^k - \Sigma_j [(f-P_j^{k+1})\zeta_i^k - P_{ij}^{k+1}]\zeta_j^{k+1}.$$

From this formula it is evident that

$$\int h_i^k(x)P(x)dx = 0 \quad \text{for all} \quad P \in \mathcal{P}_a.$$

Moreover, since $\Sigma_j \zeta_j^{k+1} = \chi_{\Omega^{k+1}}$,

$$h_i^k = \zeta_i^k f\chi_{(\Omega^{k+1})^c} - P_i^k\zeta_i^k + \zeta_i^k \Sigma_j P_j^{k+1}\zeta_j^{k+1} + \Sigma_j P_{ij}^{k+1}\zeta_j^{k+1}.$$

But $|f(x)| \leq C_1 M_{(N_p)} f(x) \leq C_1 2^{k+1}$ for almost every $x \notin \Omega^{k+1}$ by Theorem 2.6, so by Lemmas 3.8 and 3.26,

$$\|h_i^k\|_\infty \leq (2C_1 + A_2 + 2LA_2 + 2LA_{10})2^k = C_2 2^k.$$

Lastly, since $P_{ij}^{k+1} = 0$ unless $\hat{B}_i^k \cap \hat{B}_j^{k+1} \neq \emptyset$, it follows from (3.29) and Lemma 3.24 that h_i^k is supported in $B(T_2 r_i^k, x_i^k)$. Therefore, if we set

$$\lambda_i^k = C_2 2^k T_2^{Q/p}|B_i^k|^{1/p} \quad \text{and} \quad a_i^k = h_i^k/\lambda_i^k,$$

we see that a_i^k is a (p,∞,a)-atom, and that

$$\Sigma_k\Sigma_i(\lambda_i^k)^p = C_3\Sigma_k\Sigma_i 2^{kp}|B_i^k|$$

$$\leq LC_3\Sigma_k 2^{kp}|\Omega^k|$$

$$= 2LC_3\Sigma_k 2^{k(p-1)}|\Omega^k|2^{k-1}$$

$$\leq 2LC_3 p^{-1}\int_0^\infty p\alpha^{p-1}|\{x : Mf(x) > \alpha\}|d\alpha$$

$$= C_4\|Mf\|_p^p.$$

Thus the series $\Sigma_k \Sigma_i \lambda_i^k a_i^k$ converges in $H^p_{\infty,a}$ and defines an atomic decomposition of f.

It remains to remove the restriction that $f \in L^1$. If f is an arbitrary element of H^p, by Corollary 3.22 we can find a sequence $\{f_m\}$ in $H^p \cap L^1$ such that $\rho^p(f_1) \leq (3/2)\rho^p(f)$, $\rho^p(f_m) \leq 2^{-m}\rho^p(f)$ for $m > 1$, and $f = \Sigma_1^\infty f_m$. Let $f_m = \Sigma_i \lambda_m^i f_m^i$ be the atomic decomposition of f_m constructed above. Then $f = \Sigma_m \Sigma_i \lambda_m^i f_m^i$ is an atomic decomposition of f, and

$$\Sigma_m \Sigma_i (\lambda_m^i)^p \leq C_4((3/2) + \Sigma_2^\infty 2^{-m})\rho^p(f) = 2C_4\rho^p(f). \quad \#$$

(3.30) THEOREM. Suppose $0 < p \leq 1$.

(i) If (p,q,a) is admissible then $H^p = H^p_{q,a}$ and $\rho^p \sim \rho^p_{q,a}$.

(ii) If $f \in S'$ then $f \in H^p$ if and only if $M_{(N)} f \in L^p$ for some $N \geq N_p$, and $\rho^p(f) \sim \|M_{(N)} f\|_p^p$.

Proof: (i) It is easily checked that if $1 \leq q < r \leq \infty$ then every (p,r,a)-atom is also a (p,q,a)-atom, and hence $H^p_{r,a} \subset H^p_{q,a}$. Therefore by Proposition 2.15 and Theorem 3.28,

$$H^p \subset H^p_{\infty,a} \subset H^p_{q,a} \subset H^p,$$

where the inclusions are continuous.

(ii) If $f \in H^p$ then $M_{(N)} f \in L^p$ for all $N \geq N_p$ since $M_{(N)} f \leq M_{(N_p)} f$. Conversely, the proof of Theorem 3.28, with N_p replaced by N (cf. the remarks following Theorem 3.17 and Corollary 3.22), shows that if $M_{(N)} f \in L^p$ then $f \in H^p_{\infty,a}$ for any p-admissible a, and $\rho^p_{\infty,a}(f) \leq C\|M_{(N)} f\|_p^p$. Thus by (i), $f \in H^p$ and $\rho^p(f) \sim \|M_{(N)} f\|_p^p$. #

Theorem 3.30 is the rock on which our subsequent investigations will be founded, and we shall frequently use it without referring to it explicitly. As a first application, we prove that smooth functions are dense in H^p.

(3.31) LEMMA. Suppose $\phi \in S$. For any $N \in \mathbb{N}$ there exist $N' \geq N$ and $C > 0$ such that for all $f \in S'$,

$$\sup_{\varepsilon>0} M_{(N')}(f*\phi_\varepsilon) \leq CM_{(N)}f.$$

Proof: We first observe that for any $N \in \mathbb{N}$ there exist $N' \geq N$ and $C > 0$ such that for all $\phi,\psi \in S$,

$$\sup_{\delta\leq 1} \|\phi_\delta*\psi\|_{(N)} \leq C\|\phi\|_{(N')}\|\psi\|_{(N')},$$

$$\sup_{\delta\leq 1} \|\phi*\psi_\delta\|_{(N)} \leq C\|\phi\|_{(N')}\|\psi\|_{(N')}.$$

(This follows from Proposition 1.49 and its proof.) From this we see that if $f \in S'$ and $\phi,\psi \in S$,

$$\begin{aligned}\sup_{\varepsilon,t>0} |f*\phi_\varepsilon*\psi_t| &\leq \sup_{\varepsilon\leq t} |f*\phi_\varepsilon*\psi_t| + \sup_{\varepsilon>t} |f*\phi_\varepsilon*\psi_t| \\ &\leq \sup_{\varepsilon/t\leq 1} |f*(\phi_{\varepsilon/t}*\psi)_t| + \sup_{t/\varepsilon\leq 1} |f*(\phi*\psi_{t/\varepsilon})_\varepsilon| \\ &\leq 2C\|\phi\|_{(N')}\|\psi\|_{(N')}M_{(N)}f.\end{aligned}$$

Thus, regarding ϕ as fixed and ψ as varying, by Proposition 2.8 we have

$$\sup_{\varepsilon>0} M_{(N')}(f*\phi_\varepsilon) \leq C'M_{(N)}f. \quad \#$$

(3.32) COROLLARY. *If* $f \in H^p$ *and* $\phi \in \mathcal{S}$ *then* $f*\phi \in H^p$.

(3.33) THEOREM. $H^p \cap C_0^\infty$ *is dense in* H^p.

Proof: It suffices to show that if f is a finite linear combination of atoms then f can be approximated in the H^p topology by elements of C_0^∞. Choose $\phi \in C_0^\infty$ with $\int \phi = 1$. Then $f*\phi_\varepsilon \in C_0^\infty$ since f has compact support and $f*\phi_\varepsilon \in H^p$ by Corollary 3.32; we shall show that $f*\phi_\varepsilon \to f$ in H^p as $\varepsilon \to 0$. By Lemma 3.31 (with $N = N_p$) and the dominated convergence theorem it suffices to show that $M_{(N')}(f*\phi_\varepsilon - f)$ tends to zero almost everywhere. This is certainly true if $f \in C_c$: in fact, in this case,

$$\|M_{(N')}(f*\phi_\varepsilon - f)\|_\infty \leq C\|f*\phi_\varepsilon - f\|_\infty \to 0.$$

In general, since $f \in L^1$, for any $\delta > 0$ we can choose $g \in C_c$ with $\|f-g\|_1 < \delta$. Then by Lemma 3.31,

$$\limsup_{\varepsilon \to 0} M_{(N')}(f*\phi_\varepsilon - f)$$

$$\leq \sup_{\varepsilon>0} M_{(N')}((f-g)*\phi_\varepsilon) + \limsup_{\varepsilon \to 0} M_{(N')}(g*\phi_\varepsilon - g) + M_{(N')}(f-g)$$

$$\leq CM_{(N_p)}(f-g).$$

Hence by the maximal theeorem (2.4), for any $\alpha > 0$,

$$|\{x : \limsup_{\varepsilon \to 0} M_{(N')}(f*\phi_\varepsilon - f)(x) > \alpha\}| \leq C'\|f-g\|_1/\alpha < C'\delta/\alpha.$$

Since δ is arbitrary, we are done. #

Remark: C_0^∞ is not a subset of H^p for $p \leq 1$. However, if $f \in C_0^\infty$ and $\int fP = 0$ for all $P \in \mathcal{P}_a$ where a is p-admissible, f will be a constant multiple of a (p,∞,a)-atom, hence $f \in H^p$.

We conclude this section with some comments about vector-valued functions. If X is a Banach space, we can consider the space $\mathcal{S}'_X$ of X-valued tempered distributions on G, i.e., the space of continuous linear maps from $\mathcal{S}$ into X. The whole of H^p theory up to this point can be developed in this context, merely by replacing absolute values by X-norms in appropriate places.* For example, if $f \in \mathcal{S}'_X$ and $\phi \in \mathcal{S}$,

$$M_\phi f(x) = \sup_{|x^{-1}y|<t<\infty} \|f*\phi_t(y)\|_X .$$

If we define $H^p_X = \{f \in \mathcal{S}'_X : M_{(N_p)} f \in L^p\}$, all of our theorems remain valid for H^p_X, with the same proofs. Only one point may require comment. Since $\mathcal{P}_a$ is finite-dimensional, a choice of compactly supported smooth measure μ on G induces an identification of the space of linear maps from $\mathcal{P}_a$ into X with the space $(\mathcal{P}_a)_X$ of polynomials of homogeneous degree $\leq a$ with coefficients in X. Namely, if $\pi_1,\ldots,\pi_m$ is an orthonormal basis for $\mathcal{P}_a$ in $L^2(\mu)$ and $L : \mathcal{P}_a \to X$, we set

$$P_L(x) = \Sigma_1^m L(\pi_j)\overline{\pi_j(x)}.$$

Then for any $Q \in \mathcal{P}_a$, $L(Q) = \int QP_L d\mu$. Thus the construction of the polynomials

(*) However the reader should take care because not all results in H^p theory are extendable mutatis mutandis to the case of Banach space-valued functions. Some theorems require the L^2 boundedness of appropriate operators, and those may be essentially restricted to the case when the Banach space is a Hilbert space. See also the remarks preceeding theorem 6.20 below.

P_i, etc., in the Calderón-Zygmund and atomic decompositions can be carried out in the Banach space setting.

C. Interpolation Theorems

In this section we prove two results on interpolation of H^p spaces by the real method. We recall that if X and Y are quasi-normed linear spaces embedded in some topological vector space, the Peetre K-functional on $X+Y$ is defined by

$$K(t,f) = \inf\{\|g\|_X + t\|h\|_Y : g \in X,\ h \in Y,\ f = g+h\} \qquad (t > 0),$$

and the interpolation space $[X,Y]_{\theta,q}$ $(0 < \theta < 1,\ 0 < q < \infty)$ is defined by

$$[X,Y]_{\theta,q} = \{f \in X+Y : \|f\|_{\theta,q} = [\int_0^\infty (t^{-\theta}K(t,f))^q dt/t]^{1/q} < \infty\}.$$

We refer the reader to Bergh and Löfström [1] for a detailed exposition of these matters. (We remark that to fit H^p spaces $(p < 1)$ into this setting, we use the quasi-norm $(\rho^p)^{1/p}$ rather than ρ^p.)

(3.34) THEOREM. <u>If</u> $0 < p < r \leq \infty$, $0 < \theta < 1$, <u>and</u> $q^{-1} = (1-\theta)p^{-1} + \theta r^{-1}$, <u>then</u> $[H^p,H^r]_{\theta,q} = H^q$.

<u>Proof</u>: By the reiteration theorem for the interpolation functor $[\ ,\ ]_{\theta,q}$ (cf. Bergh and Löfström [1], Theorem 3.11.5), it suffices to assume that $r = \infty$. We may also assume that $p \leq 1$, since otherwise the result is well known. Thus we suppose that $0 < p \leq 1$, $0 < \theta < 1$, and $q = p/(1-\theta)$.

First, the subadditive operator $M_{(N_p)}$ maps H^p to L^p and L^∞ to L^∞ boundedly, so it maps $[H^p,L^\infty]_{\theta,q}$ to $[L^p,L^\infty]_{\theta,q}$ boundedly. But $[L^p,L^\infty]_{\theta,q} = L^q$, and $N_p \geq N_q$, whence it follows that $[H^p,L^\infty]_{\theta,q} \subset H^q$.

To prove the reverse inclusion, by Corollary 3.22 it suffices to show that $H^q \cap L^1 \subset [H^p, L^\infty]_{\theta,q}$. If $f \in H^q \cap L^1$, let F be the nonincreasing rearrangement of $Mf = M_{(N_p)}f$ on $(0,\infty)$, and for $t > 0$ let $f = g^t + \Sigma b_i^t$ be a Calderón-Zygmund decompositon of f of degree a and height $\alpha = F(t^p)$ associated to Mf, where a is p-admissible. Then by Theorem 3.17,

$$\|M(\Sigma b_i^t)\|_p^p \leq C \int_{Mf(x)>\alpha} Mf(x)^p dx = C \int_0^{t^p} F(s)^p ds.$$

Therefore, making the change of variable $t^p \to t$ and then using Hardy's inequality (cf. Stein [2], p. 272),

$$(3.35) \quad \int_0^\infty (t^{-\theta} \|M(\Sigma b_i^t)\|_p)^q dt/t \leq C \int_0^\infty t^{-\theta q} \left(\int_0^{t^p} F(s)^p ds\right)^{q/p} dt/t$$

$$= Cp^{-1} \int_0^\infty t^{-\theta q/p} \left(\int_0^t F(s)^p ds\right)^{q/p} dt/t$$

$$= Cp^{-1}\theta^{-q/p} \int_0^\infty t^{-\theta q/p} (tF(t)^p)^{q/p} dt/t$$

$$= C' \int_0^\infty F(t)^q dt$$

$$= C' \|Mf\|_q^q.$$

Moreover, by Theorem 3.20,

$$(3.36) \quad \int_0^\infty (t^{1-\theta} \|g^t\|_\infty)^q dt/t \leq C \int_0^\infty t^{(1-\theta)q} F(t^p)^q dt/t$$

$$= Cp^{-1} \int_0^\infty t^{(1-\theta)q/p} F(t)^q dt/t = Cp^{-1} \int_0^\infty F(t)^q dt$$

$$= C' \|Mf\|_q^q.$$

But clearly

$$K(t,f) \leq \|M(\Sigma b_i^t)\|_p + t\|g^t\|_\infty,$$

so by adding (3.35) and (3.36) if $q \leq 1$ or applying Minkowski's inequality to them if $q > 1$, we have

$$[\int_0^\infty (t^{-\theta}K(t,f))^q dt/t]^{1/q} \leq C\|Mf\|_q,$$

so that $f \in [H^p, L^\infty]_{\theta,q}$. #

Our second interpolation theorem is closely related to the first one, but we shall give a somewhat different proof.

(3.37) THEOREM. _Suppose_ $0 < p < 1 < q \leq \infty$, _and suppose_ T _is a subadditive operator from_ L^1 _to the space of measurable functions such that_

$$\|Tf\|_p^p \leq C\rho^p(f) \qquad \text{for } f \in L^1 \cap H^p,$$

$$\|Tf\|_q \leq C\|f\|_q \qquad \text{for } f \in L^1 \cap L^q.$$

Then T _is weak type_ $(1,1)$.

Proof: Given $f \in L^1$ and $\alpha > 0$, let $f = g + \Sigma b_i$ be a Calderón-Zygmund decomposition of f of degree a and height $\alpha/2$ associated to $M_{(0)}f$, where a is p-admissible. Then each b_i is supported in a ball $B_i (= B(2r_i, x_i)$ in the notation of Section A), so if we set $\lambda_i = \|b_i\|_1 |B_i|^{(1/p)-1}$, we see that b_i/λ_i is a $(p,1,a)$-atom. Moreover, by Hölder's inequality, Theorem 3.18, (3.4), and the maximal theorem (2.4),

$$\Sigma\lambda_i^p = \Sigma\|b_i\|_1^p|B_i|^{1-p}$$

$$\leq (\Sigma\|b_i\|_1)^p(\Sigma|B_i|)^{1-p}$$

$$\leq L^{1-p}(\Sigma\|b_i\|_1)^p|\{x : M_{(0)}f(x) > \alpha/2\}|^{1-p}$$

$$\leq C\|f\|_1^p(2\|f\|_1/\alpha)^{1-p}$$

$$\leq 2C\alpha^{p-1}\|f\|_1.$$

Hence $\Sigma b_i \in H^p$ and $\rho^p(\Sigma b_i) \leq C'\alpha^{p-1}\|f\|_1$. On the other hand, by Theorems 3.18 and 3.20, $g \in L^1 \cap L^\infty$, $\|g\|_1 \leq C\|f\|_1$, and $\|g\|_\infty \leq C\alpha$, so that

$$\|g\|_q \leq [\|g\|_\infty^{q-1} \int|g|]^{1/q} \leq C\alpha^{1-(1/q)}\|f\|_1^{1/q}.$$

Now $|Tf| \leq |Tg| + |T\Sigma b_i|$, so

$$|\{x : |Tf(x)| > \alpha\}| \leq |\{x : |Tg(x)| > \alpha/2\}| + |\{x : |T\Sigma b_i(x)| > \alpha/2\}|.$$

Therefore, by the preceding estimates,

$$|\{x : |Tf(x)| > \alpha\}| \leq (2\|Tg\|_q/\alpha)^q + (2\|T\Sigma b_i\|_p/\alpha)^p$$

$$\leq C(\|g\|_q/\alpha)^q + C\rho^p(\Sigma b_i)/\alpha^p$$

$$\leq C'[\alpha^{q-1}\|f\|_1/\alpha^q + \alpha^{p-1}\|f\|_1/\alpha^p]$$

$$= 2C'\|f\|_1/\alpha,$$

which shows that T is weak type $(1,1)$. #

Notes and References

Section A: The idea of using a smooth partition of unity, which is essentially the case $a = 0$ of the decomposition for functions, was already utilized in C. Fefferman and Stein [1] in their proof of the boundedness on $H^p(\mathbb{R}^n)$ of various singular integrals. The more refined version in which the functions b_i have higher vanishing moments (again for $L^1(\mathbb{R}^n)$) is due to Fefferman, Rivière, and Sagher [1]. In extending this construction to distributions rather than L^1 functions, we have borrowed some ideas from Macías and Segovia [2].

Section B: The history of atomic H^p spaces apparently begins with the unpublished observation of C. Fefferman that the atomic decomposition for H^1 is equivalent to the duality of H^1 and BMO. (We shall comment on this in Chapter 5.) For $0 < p \leq 1$ the atomic decompositon of H^p was proved on $\mathbb{R}$ by Coifman [1], on $\mathbb{R}^n$ by Latter [1], and on $\mathbb{R}^n$ with non-isotropic dilations by Calderón [1] and Latter and Uchiyama [1]. Coifman and Weiss [2] have developed an H^p theory on general spaces of homogeneous type by taking the atomic characterization of H^p as a definition. (In this generality, however, the theory only exists for $p_0 < p \leq 1$, where p_0 depends on the space in question and is usually strictly positive. For example, on a homogeneous group the Coifman-Weiss theory applies for $p > Q/(Q+1)$.) In the context of spaces of homogeneous type, the characterization of H^p in terms of maximal functions is due to Macías and Segovia [2]; see also Uchiyama [1]. Atomic decompositions have also been obtained for various other types of H^p spaces by Herz [1], Garnett and Latter [1], Goldberg [1], Chang and Fefferman [1], and Mauceri, Picardello, and Ricci [1].

An interesting extension of atomic H^p theory which we do not discuss in this monograph is the theory of molecules: see Coifman and Weiss [2], Coifman and Rochberg [1], Taibleson and Weiss [1], and Hemler [1].

Section C: The first interpolation theorem for $H^p(\mathbb{R}^n)$ via the real method was given by Igari [1]. For the case $G = \mathbb{R}^n$, Theorem 3.34 is due to Fefferman, Rivière, and Sagher [1], whose proof we have reproduced almost verbatim. It was extended to $\mathbb{R}^n$ with non-isotropic dilations by Calderón and Torchinsky [2], and to spaces of homogeneous type by Macías [1] (see also Coifman and Weiss [2]). It is an important separate fact that H^p spaces can be interpolated by the complex method. For the classical case of the unit disc see Zygmund [1]. The $\mathbb{R}^n$ theory was initiated in C. Fefferman and Stein [1]; some further results are in Calderón and Torchinsky [2], and Macías [1].

CHAPTER 4

Other Maximal Function Characterizations of H^p

In this chapter we investigate the relationships among various maximal functions and show that, at least when G is graded, there exist functions $\phi \in \mathcal{S}$ such that M_ϕ^0 dominates the grand maximal operator $M_{(N_p)}$ for all p. The results are as follows. First if $\phi \in \mathcal{S}$, with $\int \phi \neq 0$, is such that ϕ_t forms a commuting family under convolutions, then the maximal function formed with ϕ_t characterizes H^p. Secondly, on graded groups (those whose exponents of homogeneity can be taken to be integral) such ϕ's are shown to exist by using invariant hypoelliptic differential equations. Finally in certain special cases (which include the Heisenberg group), ϕ_t is a commuting family whenever ϕ is appropriately "radial".

A. Relationships Among Maximal Functions

We begin with the following useful result.

(4.1) THEOREM. _Suppose_ u _is a continuous function on_ $G \times (0,\infty)$. _Given_ $\alpha > 0$ _and_ $\lambda > 0$, set

$$u_\alpha^*(x) = \sup_{|x^{-1}y|<\alpha t<\infty} |u(y,t)|,$$

$$u_\lambda^{**}(x) = \sup_{y\in G,\, t>0} |u(y,t)| \left(\frac{t}{|x^{-1}y|+t}\right)^\lambda .$$

If $0 < p < \infty$ *and* $\lambda > Q/p$, *there exist* $C_1, C_2 > 0$, *depending only on* α, λ, *and* p, *such that*

$$C_1 \|u_\alpha^*\|_p \leq \|u_\lambda^{**}\|_p \leq C_2 \|u_\alpha^*\|_p .$$

Proof: Obviously $u_\alpha^*(x) \leq (1+\alpha)^\lambda u_\lambda^{**}(x)$ for every x, so we can take $C_1 = (1+\alpha)^{-\lambda}$. To prove the estimate in the other direction, set $q = Q/\lambda$, so that $q < p$. Observe that $|u(y,t)| \leq u_\alpha^*(z)$ whenever $z \in B(\alpha t, y)$ and that for any x,y, $B(\alpha t,y) \subset B(\gamma(|x^{-1}y| + \alpha t),x)$. It follows that

$$\begin{aligned} |u(y,t)|^q &\leq |B(\alpha t,y)|^{-1} \int_{B(\alpha t,y)} u_\alpha^*(z)^q dz \\ &\leq \frac{\gamma^Q(|x^{-1}y|+\alpha t)^Q}{(\alpha t)^Q} |B(\gamma(|x^{-1}y| + \alpha t),x)|^{-1} \int_{B(\gamma(|x^{-1}y|+\alpha t),x)} u_\alpha^*(z)^q \\ &\leq \frac{\gamma^Q(|x^{-1}y|+\alpha t)^Q}{(\alpha t)^Q} M_{HL}((u_\alpha^*)^q)(x) \\ &\leq C_\alpha \left(\frac{|x^{-1}y|+t}{t}\right)^Q M_{HL}((u_\alpha^*)^q)(x) \end{aligned}$$

where M_{HL} is the Hardy-Littlewood maximal function and $C_\alpha = \gamma^Q \max(\alpha^{-Q},1)$. Since $\lambda = Q/q$, this says that for all $x \in G$,

$$u_\lambda^{**}(x)^q \leq C_\alpha M_{HL}((u_\alpha^*)^q)(x),$$

and since $p/q > 1$, the maximal theorem (2.4) yields

$$\int u_\lambda^{**}(x)^p dx \leq C_\alpha^{p/q} \int [M_{HL}((u_\alpha^*)^q)(x)]^{p/q} dx \leq C \int u_\alpha^*(x)^p dx. \quad \#$$

From this we immediately deduce two things:

(1) If $f \in \mathcal{S}'$, $\phi \in \mathcal{S}$, and $\lambda > 0$, let us define the tangential maximal function $T_\phi^\lambda f$ by

$$(4.2)\qquad T_\phi^\lambda f = \sup_{y\in G,\, t>0} |f*\phi_t(y)| \left(\frac{t}{|x^{-1}y|+t}\right)^\lambda .$$

Then $M_\phi f \in L^p$ if and only if $T_\phi^\lambda f \in L^p$, provided $\lambda > Q/p$.

(2) In defining the nontangential maximal function $M_\phi f$ we could replace the cone $\{(y,t) : |x^{-1}y| < t < \infty\}$ by $\{(y,t) : |x^{-1}y| < \alpha t < \infty\}$ for any $\alpha > 0$ without changing the L^p properties.

If $p > 1$, it follows from Theorems 2.4 and 2.7 that if $\phi \in \mathcal{S}$ and $\int\phi = 1$ then

$$(4.3)\qquad \|M_\phi^0 f\|_p \sim \|M_\phi f\|_p \sim \|M_{(N_p)} f\|_p$$

for all $f \in H^p$. If $G = \mathbb{R}^n$, this result remains true for all $p > 0$. The crux of the matter is the following result, which shows how any $\psi \in \mathcal{S}$ can be expressed in terms of ϕ and its dilates:

(4.4) PROPOSITION. *If* $\phi \in \mathcal{S}(\mathbb{R}^n)$ *and* $\int\phi = 1$, *there exists* $\varepsilon > 0$ *such that any* $\psi \in \mathcal{S}$ *can be written as*

$$(4.5)\qquad \psi = \Sigma_0^\infty \phi_{\varepsilon 2^{-k}} * \psi^{(k)}, \qquad \psi^{(k)} \in \mathcal{S}.$$

Moreover, for any $m,N \in \mathbb{N}$ *there exists* $C_{m,N} > 0$, *independent of* ψ, *such that* $\|\psi^{(k)}\|_{(N)} \le C_{m,N}\, 2^{-mk} \|\psi\|_{(N)}$.

<u>Proof</u>: Taking Fourier transforms, we see that (4.5) is equivalent to

$$\hat{\psi}(\xi) = \Sigma_0^\infty \hat{\phi}(\varepsilon 2^{-k}\xi)\hat{\psi}^{(k)}(\xi). \tag{4.6}$$

Since $\hat{\phi}(0) = \int \phi = 1$, there exists $\varepsilon > 0$ such that $|\hat{\phi}(\xi)| \geq 1/2$ for $|\xi| \leq \varepsilon$, and hence $|\hat{\phi}(\varepsilon 2^{-k}\xi)| \geq 1/2$ for $|\xi| \leq 2^k$. Let $\{\zeta_k\}_0^\infty$ be a partition of unity on $\mathbb{R}^n$ such that $\operatorname{supp} \zeta_0 \subset B(1,0)$, $\operatorname{supp} \zeta_j \subset B(2^j,0)\backslash B(2^{j-2},0)$, and $\|(\partial/\partial\xi)^I \zeta_j\|_\infty \leq C2^{-j|I|}$. If we define

$$\hat{\psi}^{(k)}(\xi) = \hat{\psi}(\xi)\zeta_k(\xi)\hat{\phi}(\varepsilon 2^{-k}\xi)$$

then (4.6) obviously holds. The estimates for $\|\hat{\psi}^{(k)}\|_{(N)}$, and hence for $\|\psi^{(k)}\|_{(N)}$, follow easily from this construction; details are left to the reader. #

We do not know if this proposition remains true for an arbitrary homogeneous group G; certainly the above argument is worthless if G is non-Abelian. However, we shall now show that a variant of this proposition is true for general G when an additional restriction is imposed on ϕ, and from this we shall deduce the relations (4.3) for such ϕ's.

<u>Definition</u>: a <u>commutative</u> <u>approximate</u> <u>identity</u> is a function $\phi \in \mathcal{S}$ such that $\int \phi = 1$ and $\phi_s * \phi_t = \phi_t * \phi_s$ for all $s,t > 0$.

We postpone until Section B a discussion of the problem of finding commutative approximate identities. For the present, we assume that we have such a function in hand and proceed to work with it. If $\phi \in \mathcal{S}$, we shall denote by ϕ^{*N} the N-fold convolution product of ϕ with itself. We observe that $(\phi_s)^{*N} = (\phi^{*N})_s$ and $\int \phi^{*N} = (\int \phi)^N$.

(4.7) LEMMA. Suppose ϕ is a commutative approximate identity. If $0 < j \leq N < \infty$ and $s > 0$ then $\partial_s^j \phi_s^{*(N+1)} = \phi_s * \omega^{(j,N,s)}$ where $\partial_s = \partial/\partial s$ and $\omega^{(j,N,s)}$ is a linear combination of terms of the form $\partial_s^{j_1}\phi_s * \cdots * \partial_s^{j_N}\phi_s$ with $j_i \geq 0$ and $\Sigma_1^N j_i = j$.

Proof: Since $\partial_s^j \phi_s$ is a limit of difference quotients of ϕ_s's, the commutativity of ϕ implies that $\phi_s * (\partial_s^j \phi_s) = (\partial_s^j \phi_s) * \phi_s$ for any j. In particular, if $N \geq 1$,

(4.8) $$\partial_s \phi_s^{*(N+1)} = \Sigma_0^N \phi_s^{*i} * (\partial_s \phi_s) * \phi_s^{*(N-i)} = (N+1)\phi_s^{*N} * (\partial_s \phi_s)$$

$$= (N+1)\phi_s * [\phi_s^{*(N-1)} * \partial_s \phi_s],$$

which proves the lemma for $j = 1$. We now proceed by induction on j:

$$\partial_s^j \phi_s^{*(N+1)} = \partial_s^j(\phi_s * \phi_s^{*N}) = \Sigma_0^j \binom{j}{i} \partial_s^i \phi_s * \partial_s^{j-i} \phi_s^{*N}$$

$$= \phi_s * \partial_s^j \phi_s^{*N} + \Sigma_1^j \binom{j}{i} \partial_s^i \phi_s * \phi_s * \omega^{(j-1,N-1,s)}$$

$$= \phi_s * [\partial_s^j \phi_s^{*N} + \Sigma_1^n \binom{j}{i} \partial_s^i \phi_s * \omega^{(j-1,N-1,s)}].$$

Expanding $\partial_s^j \phi_s^{*N}$ by repeated applications of (4.8), we see that the expression in brackets has the required form. #

(4.9) THEOREM. Suppose ϕ is a commutative approximate identity. For any $\psi \in S$ and $N \in \mathbb{N}$ there exist $\theta^{(s)} \in S$, $0 < s \leq 1$, such that:

(a) $\psi = \int_0^1 \phi_s * \theta^{(s)} ds.$

(b) $\int (1+|y|)^N |\theta^{(s)}(y)|dy \leq Cs^N \|\psi\|_{(3N+3)}$, <u>where</u> C <u>depends only on</u> N <u>and</u> ϕ.

<u>Proof</u>: Fix $\zeta \in C^\infty([0,1])$ such that $\zeta(s) = s^N/N!$ for $0 \leq s \leq 1/2$, $0 \leq \zeta(s) \leq s^N/N!$ for $1/2 \leq s \leq 1$, and $\partial_s^j \zeta(1) = 0$ for $0 \leq j \leq N+1$. Also, let $\omega^{(s)} = \omega^{(N+1,N+1,s)}$ be as in Lemma 4.7. We set

$$\theta^{(s)} = (-1)^{N+1}\zeta(s)\omega^{(s)}*\psi - (\partial_s^{N+1}\zeta(s))\phi_s^{*(N+1)}*\psi,$$

and we claim that these θ's have the required properties.

(a) Consider the integral

$$I = (-1)^{N+1}\int_0^1 \zeta(s)(\partial_s^{N+1}\phi_s^{*(N+2)})*\psi ds.$$

We integrate by parts N+1 times. Because of the properties of ζ and the fact that $\partial_s^j\phi_s^{*(N+2)}*\psi$ remains bounded as $s \to 0$ (by Lemma 4.7 and Proposition 1.58), there are no boundary terms in the first N integrations by parts, and we obtain

$$I = -(\partial_s^N\zeta(s))\phi_s^{*(N+2)}*\psi\Big|_0^1 + \int_0^1 (\partial_s^{N+1}\zeta(s))\phi_s^{*(N+2)}*\psi ds$$

$$= \phi_s^{*(N+2)}*\psi\Big|_{s=0} + \int_0^1 (\partial_s^{n+1}\zeta(s))\phi_s^{*(N+2)}*\psi ds.$$

But $\int\phi^{*(N+2)} = (\int\phi)^{N+2} = 1$, so by Proposition 1.49, $\phi_s^{*(N+2)}*\psi\big|_{s=0} = \psi$. Thus by Lemma 4.7,

$$\psi = I - \int_0^1 (\partial_s^{N+1}\zeta(s))\phi_s^{*(N+2)}*\psi ds = \int_0^1 \phi_s * \theta^{(s)} ds.$$

(b) Observe that

$$|\theta^{(s)}| \le (s^N/N!)|\omega^{(s)}*\psi| + |\partial_s^{N+1}\zeta(s)||\phi^{*(N+1)}*\psi|$$

and that $|\partial_s^{N+1}\zeta(s)| \le Cs^N$ since $\partial_s^{N+1}\zeta(s) = 0$ for $s \le 1/2$. Hence we need only show that

$$\int (1+|y|)^N|\omega^{(s)} \psi(y)|dy \le C\|\psi\|_{(3N+3)},$$

$$\int (1+|y|)^N|\phi_s^{*(N+2)}*\psi(y)|dy \le C\|\psi\|_{(3N+3)}.$$

But this follows from Proposition 1.55. #

(4.10) THEOREM. Suppose ϕ is a commutative approximate identity. For any $N \in \mathbb{N}$ there exists $C > 0$ such that for all $f \in S'$ and $x \in G$, $M_{(3N+3)}f(x) \le CT_\phi^N f(x)$, where T_ϕ^N is defined by (4.2).

Proof: Suppose $\psi \in S$ and $\|\psi\|_{(3N+3)} \le 1$. Write $\psi = \int_0^1 \phi_s*\theta^{(s)}ds$ as in Theorem 4.9. Then for any $x,y \in G$,

$$|f*\phi_t(y)| = |\int_0^1 f*\phi_{st}*\theta_t^{(s)}(y)ds|$$

$$\le \int_0^1\int_G |f*\phi_{st}(yz^{-1})||\theta^{(s)}(z/t)|t^{-Q}dzds$$

$$\le \int_0^1\int_G T_\phi^N f(x)\left(\frac{|x^{-1}yz^{-1}|+st}{st}\right)^N|\theta^{(s)}(z/t)|t^{-Q}dzds$$

$$= T_\phi^N f(x)\int_0^1\int_G s^{-N}\left(\left|\frac{x^{-1}y}{t}w^{-1}\right|+1\right)^N|\theta^{(s)}(w)|dwds.$$

But if $|x^{-1}y| < t$, then

$$\left|\frac{x^{-1}y}{t} w^{-1}\right| + 1 \leq \gamma(1+|w|) + 1 \leq 2\gamma(1+|w|).$$

Therefore, by Theorem 4.9,

$$M_\psi f(x) \leq (2\gamma)^N T_\phi^N f(x) \int_0^1 \int_G s^{-N}(1+|w|)^N |\theta^{(s)}(w)| dw ds$$

$$\leq C T_\phi^N f(x),$$

from which the desired result is immediate. #

(4.11) COROLLARY. Suppose ϕ is a commutative approximate identity, $f \in S'$, and $0 < p \leq \infty$. If $M_\phi f \in L^p$ then $f \in H^p$, and there is a constant C, independent of f, such that $\rho^p(f) \leq C\|M_\phi f\|_p^p$.

Proof: Apply Theorems 4.1 and 3.30.

Having established this, we now prove that M_ϕ can be replace by the radial maximal operator M_ϕ^0.

(4.12) THEOREM. Suppose ϕ is a commutative approximate identity, $f \in S'$, and $0 < p \leq \infty$. If $M_\phi^0 f \in L^p$ then $M_\phi f \in L^p$, and there is a constant C, independent of f, such that $\|M_\phi f\|_p \leq C\|M_\phi^0 f\|_p$.

For the proof we need some preliminary arguments. If $f \in S'$, $\psi \in S$, $K \in \mathbb{N}$, $N \in \mathbb{N}$, $0 < \alpha < \infty$, and $0 < \varepsilon \leq 1$, let us set

$$M_\psi^{\varepsilon\alpha K} f(x) = \sup_{|x^{-1}y| < \alpha t < \alpha/\varepsilon} |f*\psi_t(y)| \left(\frac{t}{t+\varepsilon}\right)^K (1+\varepsilon|y|)^{-K},$$

$$T_\psi^{\varepsilon N K} f(x) = \sup_{y \in G, t < 1/\varepsilon} |f*\psi_t(y)| \left(\frac{t}{|x^{-1}y|+t}\right)^N \left(\frac{t}{t+\varepsilon}\right)^K (1+\varepsilon|y|)^{-K}.$$

(4.13) LEMMA. For each $\alpha > 0$, $p > 0$, $N > Q/p$ and $\psi \in \mathcal{S}$ there exists $C > 0$ such that for all $f \in \mathcal{S}'$, $\varepsilon \in (0,1]$, and $K \in \mathbb{N}$,

$$\|T_{\psi}^{\varepsilon NK} f\|_p \leq C\|M_{\psi}^{\varepsilon\alpha K} f\|_p.$$

Proof: Same as the proof of Theorem 4.1.

(4.14) LEMMA. For any $\alpha > 0$, $p > 0$, $f \in \mathcal{S}'$, and $\psi \in \mathcal{S}$ there exists $K \in \mathbb{N}$ such that $M_{\psi}^{\varepsilon\alpha K} f \in L^p \cap L^{\infty}$ for $0 < \varepsilon \leq 1$.

Proof: By Proposition 1.48 there exists $L \in \mathbb{N}$ and $C > 0$ (depending only on f) such that

$$|f*\psi_t(y)| \leq C\|\psi_t\|_{(L)}(1 + |y|)^L.$$

Now,

$$\|\psi_t\|_{(L)} = \sup_{y\in G, |I|\leq L} (1 + |y|)^{(Q+1)(L+1)} t^{-Q-d(I)} |Y^I\psi(y/t)|$$

$$= \sup_{y\in G, |I|\leq L} (1 + |ty|)^{(Q+1)(L+1)} t^{-Q-d(I)} |Y^I\psi(y)|.$$

Thus if $t \leq 1$,

$$\|\psi_t\|_{(L)} \leq t^{-Q(L+1)} \sup_{y,I} (1 + |y|)^{(Q+1)(L+1)} |Y^I\psi(y)|$$

$$\leq t^{-(Q+1)(L+1)} \|\psi\|_{(L)},$$

while if $1 \leq t \leq 1/\varepsilon$,

$$\|\psi_t\|_{(L)} \leq t^{-Q} t^{(Q+1)(L+1)} \sup_{y,I} (1 + |y|)^{(Q+1)(L+1)} |Y^I\psi(y)|$$

$$\leq \varepsilon^{-(Q+1)(L+1)} \|\psi\|_{(L)}.$$

Now choose $K \in \mathbb{N}$ so large that $K > L + (Q/p)$ and $K \geq (Q+1)(L+1)$. then for $t \leq 1$,

$$|f*\psi_t(y)|(\tfrac{t}{t+\varepsilon})^K(1+\varepsilon|y|)^{-K}$$

$$\leq Ct^{-(Q+1)(L+1)}\|\psi\|_{(L)}(1+|y|)^L(\tfrac{t}{t+\varepsilon})^K(1+\varepsilon|y|)^{-K}$$

$$\leq C\varepsilon^{-2K}\|\psi\|_{(L)}(1+|y|)^{L-K},$$

while for $1 \leq t \leq 1/\varepsilon$,

$$|f*\psi_t(y)|(\tfrac{t}{t+\varepsilon})^K(1+\varepsilon|y|)^{-K}$$

$$\leq C\varepsilon^{-(Q+1)(L+1)}\|\psi\|_{(L)}(1+|y|)^L(\tfrac{t}{t+\varepsilon})^K(1+\varepsilon|y|)^{-K}$$

$$\leq C\varepsilon^{-2K}\|\psi\|_L(1+|y|)^{L-K}.$$

Thus there exists C' (depending on f,ψ,ε) such that

$$M_\psi^{\varepsilon\alpha K}f(x) \leq C' \sup_{|x^{-1}y|<\alpha/\varepsilon} (1+|y|)^{L-K}$$

$$= C' \sup_{|y|<\alpha/\varepsilon} (1+|xy|)^{L-K}$$

$$\leq C'\gamma^{K-L} \sup_{|y|<\alpha/\varepsilon} (1+|x|)^{L-K}(1+|y|)^{K-L}$$

$$\leq C''(1+|x|)^{L-K},$$

where we have used Lemma 1.10. Since $L-K < -Q/p$, we are done. #

(4.15) LEMMA. Suppose ϕ is a commutative approximate identity. For any $\alpha > 0$, $N \in \mathbb{N}$, and $K \in \mathbb{N}$ there exists $C > 0$ such that for all $f \in S'$, $\varepsilon \in (0,1]$, $x \in G$, and $j = 1,\ldots,n$,

$$M_{X_j\phi}^{\varepsilon\alpha K} f(x) \le C T_\phi^{\varepsilon NK} f(x).$$

Proof: The argument is essentially the same as the proof of Theorem 4.10. Let us fix j. By Theorem 4.9 we can find $\theta^{(s)} \in S$, $0 < s \le 1$, such that

$$X_j\phi = \int_0^1 \phi_s * \theta^{(s)} ds,$$

$$\int (1 + |y|)^{N+K} |\theta^{(s)}(y)| dy \le Cs^{N+K},$$

where C Depends only on N,K, and ϕ. Then if $x,y \in G$,

$$|f*(X_j\phi)_t(y)|$$

$$\le \int_0^1 \int_G |f*\phi_{st}(yz^{-1})| |\theta^{(s)}(s/t)| t^{-Q} dz ds$$

$$\le T_\phi^{\varepsilon NK} f(x) \int_0^1 \int_G \left(\frac{|x^{-1}yz^{-1}| + st}{st}\right)^N \left(\frac{st}{st+\varepsilon}\right)^{-K} (1 + \varepsilon |yz^{-1}|)^K |\theta^{(s)}(z/t)| t^{-Q} dz ds$$

$$\le T_\phi^{\varepsilon NK} f(x) \left(\frac{t}{t+\varepsilon}\right)^{-K} \int_0^1 \int_G s^{-N-K} \left(\left|\frac{x^{-1}y}{t} w^{-1}\right| + 1\right)^N (1 + \varepsilon |y(tw^{-1})|)^K |\theta^{(s)}(w)| dw ds.$$

But if $|x^{-1}y| < \alpha t < \alpha/\varepsilon$ we have

$$\left|\frac{x^{-1}y}{t} w^{-1}\right| + 1 \le \gamma(\alpha + |w|) + 1 \le \gamma(\alpha + 1)(1 + |w|),$$

$$(1 + \varepsilon |y(tw^{-1})|)^K \le \gamma^K (1 + \varepsilon t |w|)^K (1 + \varepsilon |y|)^K \le \gamma^K (1 + |w|)^K (1 + \varepsilon |y|)^K,$$

so it follows that

$$M_{X_j,\phi}^{\varepsilon\alpha K}f(x) \le \gamma^{N+K}(\alpha+1)^N T_\phi^{\varepsilon NK}f(x) \int_0^1 \int_G s^{-N-K}(1+|w|)^{N+K}|\theta^{(s)}(w)|\,dwds$$

$$\le CT_\phi^{\varepsilon NK}f(x). \quad \#$$

<u>Proof of Theorem 4.12</u>: Given $f \in S'$ such that $M_\phi^0 f \in L^p$, choose K by Lemma 4.14 so large that $M_\phi^{\varepsilon 1K} f \in L^p \cap L^\infty$ for $0 < \varepsilon \le 1$. With this choice of K, let us set

$$f_\varepsilon^* = M_\phi^{\varepsilon 1K} f, \qquad F_\varepsilon^* = \sup_{1 \le j \le n} M_{X_j,\phi}^{\varepsilon(2\gamma)K} f.$$

By Lemmas 4.13 and 4.15, we have $F_\varepsilon^* \in L^p$ and $\|F_\varepsilon^*\|_p \le C_1 \|f_\varepsilon^*\|_p$, where C_1 is independent of ε. Let $C_2 = 2^{1/p}C_1$, and set

$$\Omega_\varepsilon = \{x : F_\varepsilon^*(x) \le C_2 f_\varepsilon^*(x)\}.$$

Then

$$\int_{\Omega_\varepsilon^c} f_\varepsilon^*(x)^p dx \le \int_{\Omega_\varepsilon^c} [C_2^{-1}F_\varepsilon^*(x)]^p dx \le C_2^{-p}C_1^p \int f_\varepsilon^*(x)^p dx$$

$$\le \frac{1}{2}\int f_\varepsilon^*(x)^p dx,$$

so that

$$\int f_\varepsilon^*(x)^p dx \le 2\int_{\Omega_\varepsilon} f_\varepsilon^*(x)^p dx.$$

<u>Claim</u>: If $0 < r < p$, there exists $C_3 > 0$, independent of ε, such that $f^*_\varepsilon(x) \leq C_3\mu(x)$ for all $x \in \Omega_\varepsilon$, where

$$\mu(x) = [M_{HL}((M^0_\phi f)^r)(x)]^{1/r}.$$

Assuming this for the moment, we complete the proof. Fix $r < p$: then by the maximal theorem (2.4),

$$\int f^*_\varepsilon(x)^p dx \leq 2\int_{\Omega_\varepsilon} f^*_\varepsilon(x)^p dx \leq 2C_3^p \int_{\Omega_\varepsilon} \mu(x)^p dx$$

$$\leq 2C_3^p \int [M_{HL}((M^0_\phi f)^r)(x)]^{p/r} dx$$

$$\leq C_4 \int M^0_\phi f(x)^p dx,$$

where C_4 depends only on C_3 and p/r. Since f^*_ε increases pointwise to $M_\phi f$ as $\varepsilon \to 0$, the monotone convergence theorem implies that $M_\phi f \in L^p$. It is not immediately clear that the final constant C_4 is independent of f (it depends on K, which depends on f). However, once we know that $M_\phi f \in L^p$ we can take $K = 0$, and the same argument then yields $\|M_\phi f\|_p \leq C\|M^0_\phi f\|_p$ with C independent of f.

It therefore remains to prove the claim. Fix $x \in \Omega_\varepsilon$ and fix $y \in G$, $t > 0$ such that $|x^{-1}y| < t < 1/\varepsilon$ and

$$|f*\phi_t(y)|(\tfrac{t}{t+\varepsilon})^K(1+\varepsilon|y|)^{-K} \geq \tfrac{1}{2}f^*_\varepsilon(x).$$

By definition of Ω_ε, since $(X_j\phi)_t = t^{d_j}X_j(\phi_t)$, for all z with

$|x^{-1}z| < 2\gamma t$ we have

$$\sup_j t^{d_j}|X_j(f*\phi_t)(z)|(1+\varepsilon|z|)^{-K} \leq C_2(\tfrac{t}{t+\varepsilon})^{-K} f_\varepsilon^*(x)$$

$$\leq 2C_2|f*\phi_t(y)|(1+\varepsilon|y|)^{-K}.$$

Since

$$\varepsilon|z| \leq \gamma\varepsilon(|x^{-1}z| + |x|) \leq \gamma\varepsilon(|x^{-1}z| + \gamma(|y^{-1}x| + |y|))$$

$$\leq \gamma\varepsilon(2\gamma\varepsilon^{-1} + \gamma(\varepsilon^{-1} + |y|)) \leq \gamma^2(3+\varepsilon|y|),$$

and similarly

$$\varepsilon|y| \leq \gamma(3+\varepsilon|z|),$$

the ratio $(1+\varepsilon|z|)/(1+\varepsilon|y|)$ is bounded above and below independently of ε, so there exists C_5 such that

$$(4.16) \quad \sup_j t^{d_j}|X_j(f*\phi_t)(z)| \geq \tfrac{1}{2}|f*\phi_t(y)| \qquad \text{for } |x^{-1}z| < 2\gamma t.$$

Next, let $\overline{C},\beta$ be as in the mean value theorem (1.33), and let $C_6 = \max(\beta, 2n\overline{C}C_5)$. We assert that

$$|f*\phi_t(w)| \geq \tfrac{1}{2}|f*\phi_t(y)| \qquad \text{for } w \in B(t/C_6, y).$$

Indeed, if $|y^{-1}w| < t/C_6$ and $|y^{-1}z| < \beta|y^{-1}w|$ then $|y^{-1}z| < t$ and hence $|x^{-1}z| < 2\gamma t$. Therefore, setting $g(u) = f*\phi_t(u^{-1})$, by the mean value theorem (1.33) and (4.16) we have

$$|f*\phi_t(w) - f*\phi_t(y)| = |g(w^{-1}) - g(y^{-1})|$$

$$\leq \overline{C}\Sigma_1^n |w^{-1}y|^{d_j} \sup_{|z^{-1}y| \leq \beta|w^{-1}y|} |Y_j g(z^{-1})|$$

$$\leq \overline{C}\Sigma_1^n |w^{-1}y|^{d_j} \sup_{|x^{-1}z| \leq 2\gamma t} |Y_j g(z^{-1})|$$

$$= \overline{C}\Sigma_1^n |y^{-1}w|^{d_j} \sup_{|x^{-1}z| \leq 2\gamma t} |X_j(f*\phi_t)(z)|$$

$$\leq \overline{C}C_5\Sigma_1^n (|y^{-1}w|/t)^{d_j} \, |f*\phi_t(y)|$$

$$\leq \overline{C}C_5\Sigma_1^n C_6^{-d_j} |f*\phi_t(y)|$$

$$\leq \tfrac{1}{2}|f*\phi_t(y)|,$$

so that $|f*\phi_t(w)| \geq \frac{1}{2}|f*\phi_t(y)|$ as desired. In particular, if $w \in B(t/C_6, y)$,

$$|f*\phi_t(w)| \geq \tfrac{1}{2}|f*\phi_t(y)| \geq \tfrac{1}{4} f_\varepsilon^*(x).$$

Also, since $C_6 \geq 1$, we have $B(t/C_6, y) \subset B(2\gamma t, x)$. Therefore,

$$\mu(x)^r \geq |B(2\gamma t, x)|^{-1} \int_{B(2\gamma t, x)} M_\phi^0 f(w)^r dw$$

$$\geq |B(2\gamma t, x)|^{-1} \int_{B(2\gamma t, x)} |f*\phi_t(w)|^r dw$$

$$\geq |B(2\gamma t, x)|^{-1} \int_{B(t/C_6, y)} |f*\phi_t(w)|^r dw$$

$$\geq (2\gamma C_6)^{-Q} 4^{-r} f_\varepsilon^*(x)^r.$$

This establishes the claim and thus completes the proof of the theorem. #

Combining this with Corollary 4.11, we obtain the final result:

(4.17) COROLLARY. Suppose ϕ is a commutative approximate identity. If $f \in S'$ and $0 < p \leq \infty$, the following are equivalent:

(a) $M_\phi^0 f \in L^p$, (b) $M_\phi f \in L^p$, (c) $M_\psi f \in L^p$ for all $\psi \in S$, (d) $f \in H^p$.

Moreover $\rho^p(f) \sim \|M_\phi^0 f\|_p$.

B. Construction of Commutative Approximate Identities

On a general homogeneous group the existence of commutative approximate identities is, as far as we know, an open question. However, on a stratified group, one example comes immediately to mind. Namely, let h be the heat kernel discussed in Chapter 1, Section G and let $\phi(x) = h(x,1)$. Then by Propositions 1.68 and 1.74 we have $\phi \in S$, $\int \phi = 1$, and $\phi_t(x) = h(x,t^2)$, hence

$$\phi_t * \phi_s = h(x,t^2+s^2) = \phi_s * \phi_t .$$

Thus ϕ is a commutative approximate identity.

We shall present a generalization of this construction which works on any graded group. Our argument depends on Helffer and Nourrigat's solution of the Rockland conjecture, which we now describe.*

*The conjecture is in Rockland [1]. The necessity had been stated and essentially proved in Rothschild and Stein [1], pp. 257-261.

Let G be a graded group, and let $\hat{G}$ denote the set of irreducible unitary representations of G. (By abuse of language, we identify a representation with its unitary equivalence class.) If $\pi \in \hat{G}$, let X_π be the Hilbert space on which π acts and let $S_\pi \subset X_\pi$ be the space of C^∞ vectors for π. (In the standard Kirillov form of π, $X_\pi = L^2(\mathbb{R}^k)$ for a suitable k and $S_\pi = S(\mathbb{R}^k)$.) Then π determines a representation $d\pi$ of $\mathfrak{g}$ as skew-Hermitian operators on S_π, which extends to a representation (still denoted by $d\pi$) of the algebra of left-invariant differential operators on G as operators on S_π.

We shall call a differential operator L on G a <u>Rockland operator</u>, or R-<u>operator</u>, if (i) L is left-invariant and homogeneous, and (ii) $d\pi(L)$ is injective on S_π for every $\pi \in \hat{G}$ except the trivial representation. The theorem of Helffer and Nourrigat [1] is then the following:

(4.18) PROPOSITION. *If L is an R-operator on a graded group G, then L is hypoelliptic (that is, if f is a distribution on G such that Lf is C^∞ on an open $\Omega \subset G$, then f is C^∞ on Ω). Moreover, if D is the degree of homogeneity of L, then $L + \Sigma_{d(I)<D}\, a_I X^I$ is hypoelliptic for any constants a_I.*

We shall be interested in R-operators L which are <u>positive</u>, that is, which are formally self-adjoint and satisfy $\int (L\phi)\bar{\phi} \geq 0$ for all $\phi \in S$. Let us make a couple of elementary observations concerning such operators:

(4.19) If L is a positive R-operator, then so is L^m for any positive integer m.

(4.20) A left-invariant differential operator L which is homogeneous of degree D is positive provided L is of the form

$$L = \Sigma_{d(I)=D,\ |I|\ \text{even}} (-1)^{|I|/2} a_I X^I, \qquad a_I \geq 0.$$

(This is clear since the vector fields X_i are formally skew-adjoint.) In this case, $d\pi(L)$ is also a positive operator for all $\pi \in \hat{G}$.

It is easy to construct examples of positive R-operators on any graded group. For instance, choose $M \in \mathbb{N}$ such that $M/4d_i$ is an integer for $i = 1,\ldots,n$ (which is possible since the exponents d_i are all rational), and set $L = \Sigma_1^n X_i^{M/d_i}$. Then L is positive by (4.20). To verify the Rockland condition, suppose $\pi \in \hat{G}$, $v \in S_\pi$, and $d\pi(L)v = 0$. If $(\ ,\ )$ and $\|\ \|$ denote the scalar product and norm on X_π, we have

$$0 = (d\pi(L)v,v) = \Sigma \| d\pi(X_i)^{M/2d_i} v\|^2.$$

Hence $d\pi(X_i)^{M/2d_i} v = 0$ for all i, so

$$0 = (d\pi(X_i)^{M/2d_i} v,v) = \pm \| d\pi(X_i)^{M/4d_i} v\|^2.$$

If $M/4d_i$ is even, the same argument shows that $d\pi(X_i)^{M/8d_i} v = 0$. If $M/4d_i$ is odd, we have $d\pi(X_i)^{(M/4d_i)+1} v = 0$, and hence as before, $d\pi(X_i)^{(M/8d_i)+(1/2)} v = 0$. Continuing inductively, we eventually obtain $d\pi(X_i)v = 0$ for all i, which implies that either $v = 0$ or π is the trivial representation.

For the remainder of this discussion we fix a positive R-operator L which is homogeneous of degree D. We make $G \times \mathbb{R}$ into a graded group by means of the dilations

$$\delta_r(x,t) = (rx, r^D t).$$

Thus if ∂_t denotes the coordinate vector field on $\mathbb{R}$, regarded as a differential operator on $G \times \mathbb{R}$, ∂_t is homogeneous of degree D.

(4.21) LEMMA. $L + \partial_t$ is an R-operator on $G \times \mathbb{R}$.

Proof: $(G \times \mathbb{R})^\wedge$ is easily seen to be isomorphic to $\hat{G} \times \mathbb{R}$: namely, if $\pi \in \hat{G}$ and $\lambda \in \mathbb{R}$, the corresponding representation $\rho = \rho_{\pi,\lambda}$ of $G \times \mathbb{R}$ is given by $\rho(x,t) = e^{i\lambda t}\pi(x)$ on the Hilbert space X_π, and $S_\rho = S_\pi$. If $v \in S_\pi$, $v \neq 0$, and $d\rho(L + \partial_t)v = 0$, we have

$$0 = (d\rho(L + \partial_t)v, v) = (d\pi(L)v, v) + i\lambda\|v\|^2.$$

In view of (4.20), $(d\pi(L)v,v) \geq 0$, so by taking real and imaginary parts we obtain

$$(d\pi(L)v,v) = \lambda\|v\|^2 = 0.$$

This implies, first, that $\lambda = 0$, and second, since the quadratic form $u \to (d\pi(L)u,u)$ is nonnegative, that $d\pi(L)v = 0$, whence π is trivial. Thus ρ is the trivial representation of $G \times \mathbb{R}$. #

We propose to construct a commutative approximate identity out of a fundamental solution for $L + \partial_t$. To begin with, by (4.19), $L^m | C_0^\infty$ (regarded as a densely defined operator on L^2) is a positive Hermitian operator for any positive integer m. By a theorem of Nelson and Stinespring [1], $L^m | C_0^\infty$ is essentially self-adjoint. (Nelson and Stinespring state their theorem for elliptic operators, but the proof uses only the hypoellipticity of $L^m + 1$, which follows from Proposition 4.18.) We denote by $\bar{L}^m$ the unique self-adjoint extension of $L^m | C_0^\infty$, noting that $\bar{L}^m = (\bar{L})^m$.

$\bar{L}$ is a positive self-adjoint operator, so it generates a contraction semigroup $\{A_t : t > 0\}$ on L^2: namely, if $\int_0^\infty \lambda dE(\lambda)$ is the spectral resolution of L,

$$A_t = \exp(-t\bar{L}) = \int_0^\infty e^{-\lambda t} dE(\lambda).$$

By the Schwartz kernel theorem (cf. Trèves [1], Chapter 51), for each $t > 0$ there is a tempered distribution $K(\cdot,t)$ on G such that

$$A_t u = u * K(\cdot,t) \qquad (u \in S).$$

Moreover, K is a continuous function from $(0,\infty)$ to S' and hence defines a distribution on $G \times (0,\infty)$. Since A_t is self-adjoint and commutes with complex conjugation, $K(\cdot,t)$ is symmetric about the origin, that is, $K(x,t) = K(x^{-1},t)$.

(4.22) LEMMA. (a) $(L + \partial_t)K = 0$ _in the sense of distributions on_ $G \times (0,\infty)$.

(b) K _is_ C^∞ _on_ $G \times (0,\infty)$.

(c) _For all_ $x \in G$, $t > 0$, _and_ $r > 0$, $K(rx, r^D t) = r^{-Q} K(x,t)$.

Proof: The proof of (a) and (c) is the same as the proof of parts (i) and (iv) of Proposition 1.68, and (b) follows from (a) by Proposition 4.18 and Lemma 4.21. #

(4.23) LEMMA. $K(\cdot,t) \in L^2$ <u>for all</u> $t > 0$, <u>and</u>

$$\int |K(x,t)|^2 dx = t^{-Q/D} \int |K(x,1)|^2 dx.$$

<u>Proof</u>: If $u \in L^2$, let $Tu(x,t) = (A_t u)(x)$. Clearly T is continuous from L^2 to the space of distributions on $G \times (0,\infty)$. However, the range of T lies in the nullspace of $L + \partial_t$, on which the distribution topology coincides with the C^∞ topology (cf. Trèves [1], Chapter 52), so the linear functional $u \to Tu(0,1)$ is bounded on L^2. Moreover, $Tu(0,1) = \int u(x)K(x,1)dx$ for $u \in S$, so by the converse to the Schwarz inequality, $K(\cdot,1) \in L^2$. The desired result now follows from Lemma 4.22(c). #

We know that K is a smooth function on $G \times (0,\infty)$. We now extend it to $G \times \mathbb{R}$ by setting $K(x,t) = 0$ for $t \leq 0$.

(4.24) LEMMA. <u>Suppose</u> $D > Q/2$. <u>Then</u> K <u>is locally integrable on</u> $G \times \mathbb{R}$ <u>and</u> $(L + \partial_t)K = \delta$ <u>where</u> δ <u>is the point mass at</u> $(0,0) \in G \times \mathbb{R}$.

(Remark: we shall show later that the hypothesis $D > Q/2$ is superfluous.)

<u>Proof</u>: Clearly K is locally integrable except perhaps near $t = 0$. However, by Lemma 4.23, for any $\varepsilon > 0$, $R > 0$,

$$\int_0^\varepsilon \int_{|x|<R} |K(x,t)|dxdt \leq \int_0^\varepsilon [\int_G |K(x,t)|^2 dx]^{1/2} [\int_{|x|<R} dx]^{1/2} dt$$

$$= R^{Q/2} [\int_G |K(x,1)|^2 dx]^{1/2} \int_0^\varepsilon t^{-Q/2D} dt$$

$$= CR^{Q/2} \varepsilon^{1-(Q/2D)}$$

since we assumed $D > Q/2$. This shows that G is locally integrable on $G \times \mathbb{R}$ and hence defines a distribution. Moreover, if we set $K^{\varepsilon}(x,t) = K(x,t)$ for $t > \varepsilon$, $K^{\varepsilon}(x,t) = 0$ for $t \leq \varepsilon$, this estimate also shows that $K^{\varepsilon} \to K$ in the distribution topology as $\varepsilon \to 0$. Hence, to prove that $(\mathcal{L} + \partial_t)K = \delta$ it suffices to show that $(\mathcal{L} + \partial_t)K^{\varepsilon} \to \delta$ in the distribution topology as $\varepsilon \to 0$, and this is equivalent to the assertion that for all $u \in C_0^{\infty}(G \times \mathbb{R})$, $(\mathcal{L} + \partial_t)(u*K^{\varepsilon}) \to u$ pointwise as $\varepsilon \to 0$ (convolution on $G \times \mathbb{R}$).

To establish this, observe that by Lemma 4.22(a,b),

$$(\mathcal{L} + \partial_t)(u*K^{\varepsilon})(x,t) = (\mathcal{L} + \partial_t) \int_{-\infty}^{t-\varepsilon} \int_G u(y,s)K(y^{-1}x,t-s)dyds$$

$$= \int_G u(y,t-\varepsilon)K(y^{-1}x,\varepsilon)dy.$$

Fix t and set $u_{\varepsilon}(y) = u(y,t-\varepsilon)$ and $K_{\varepsilon}(y) = K(y,\varepsilon)$. Then the last integral is $u_{\varepsilon}*K_{\varepsilon}(x)$ (convolution on G), and we have

$$u_{\varepsilon}*K_{\varepsilon}-u_0 = (u_{\varepsilon}-u_0)*K_{\varepsilon} + (u_0*K_{\varepsilon}-u_0).$$

On the one hand, since $u \in C_0^{\infty}$ and $D > Q/2$, by Lemma 4.23 we have

$$\|(u_{\varepsilon}-u_0)*K_{\varepsilon}\|_{\infty} \leq \|u_{\varepsilon}-u_0\|_2 \, \|K_{\varepsilon}\|_2 \leq C\varepsilon \cdot \varepsilon^{-Q/2D} \to 0 \qquad \text{as} \quad \varepsilon \to 0.$$

On the other hand,

$$u_0*K_{\varepsilon} = A_{\varepsilon}u_0 \to u_0 \quad \text{in} \quad L^2 \quad \text{as} \quad \varepsilon \to 0,$$

and since $\mathcal{L}^m$ commutes with A_{ε} on $\mathrm{Dom}(\overline{\mathcal{L}}^m)$,

$$\mathcal{L}^m(u_0*K_{\varepsilon}) = A_{\varepsilon}\mathcal{L}^m u_0 \to \mathcal{L}^m u_0 \quad \text{in} \quad L^2 \quad \text{as} \quad \varepsilon \to 0.$$

Therefore, to finish the proof it suffices to establish the following:

Claim: If m is a sufficiently large integer then the elements of $\mathrm{Dom}(\bar{L}^m)$ are continuous, and for each compact $\Omega \subset G$ there is a constant C such that

$$\sup_{y\in\Omega} |v(y)| \leq C(\|v\|_2 + \|L^m v\|_2) \quad \text{for all} \quad v \in \mathrm{Dom}(\bar{L}^m).$$

Helffer and Nourrigat [1] (Proposition 6.4) show that if m is sufficiently large, we have the estimate

$$\Sigma_{d(I)\leq mD} \|X^I v\|_2 \leq C(\|v\|_2 + \|L^m v\|_2)$$

for all $v \in C_0^\infty$ and hence (by the result of Nelson and Stinespring quoted above) for all $v \in \mathrm{Dom}(\bar{L}^m)$. Pick an integer $r > (\dim G)/2$. By increasing m we may assume that $d(I) \leq mD$ whenever $|I| \leq r$, so that

$$\Sigma_{|I|\leq r} \|X^I v\|_2 \leq C(\|v\|_2 + \|L^m v\|_2) \quad \text{for} \quad v \in \mathrm{Dom}(\bar{L}^m).$$

The claim now follows from the classical Sobolev imbedding theorem (cf. Stein [2], p. 124). #

(4.25) THEOREM. Let L be a positive R-operator on the graded group G, and let $\phi = K(\cdot,1)$ be the distribution kernel of $A_1 = \exp(-\bar{L})$. Then $\phi \in S$, and ϕ is a commutative approximate identity.

Proof: First suppose that L is homogeneous of degree $D > Q/2$. By Lemmas 4.21 and 4.24, K is C^∞ on $(G\times\mathbb{R})\setminus\{(0,0)\}$. Since $K(x,t) = 0$ for $t \leq 0$, this implies that for $x \neq 0$, $K(x,t)$ and all of its x-derivatives vanish to infinite order as $t \to 0$. Therefore, by the homogeneity of K,

$K(x,t)$ and all of its x-derivatives vanish to infinite order as $x \to \infty$ (cf. the proof of Proposition 1.74), that is, $K(\cdot,t) \in \mathcal{S}$.

At this point we can remove the restriction that $D > Q/2$ by using the "principle of subordination". If L is any positive R-operator, homogeneous of degree D, and m is a positive integer, let $A_t^{(m)} = \exp(-t\bar{L}^{2^m})$. Then

$$(4.26) \qquad A_t^{(m-1)} = \int_0^\infty \frac{e^{-s}}{\sqrt{\pi s}} A^{(m)}_{t^2/4s}\, ds.$$

(This follows by applying the functional calculus to the scalar equation

$$e^{-\lambda t} = \int_0^\infty \frac{e^{-s}}{\sqrt{\pi s}} e^{-\lambda^2 t^2/4s}\, ds,$$

for which see Stein [2], p. 61.) Let us choose m large enough so that $2^m D > Q/2$, and for $j \leq m$ let $K^{(j)}(\cdot,t)$ be the distribution kernel of $A_t^{(j)}$, which is a smooth function by Lemma 4.22. By the preceding arguments with L replaced by L^{2^m}, we have $K^{(m)}(\cdot,t) \in \mathcal{S}$, and thus by (4.26),

$$K^{(m-1)}(x,t) = \int_0^\infty \frac{e^{-s}}{\sqrt{\pi s}} K^{(m)}(x,t^2/4s)\, ds,$$

where the integral converges since $K^{(m)}(x,t^2/4s)$ vanishes as $s \to 0$ and grows at most like $s^{Q/2^m D}$ as $s \to \infty$ (even for $x = 0$). Moreover, by Lemma 4.22(c), $\int |K^{(m)}(x,t)|dx$ is independent of t. Therefore,

$$\int |K^{(m-1)}(x,t)|dx \leq \int_G \int_0^\infty \frac{e^{-s}}{\sqrt{\pi s}} |K^{(m)}(x,t^2/4s)|\, ds dx$$

$$= C \int_0^\infty \frac{e^{-s}}{\sqrt{\pi s}}\, ds = C,$$

so that $K^{(m-1)}(\cdot,t) \in L^1$. Once this is known, the arguments used to prove Propositions 1.71 and 1.74 show that $(\mathcal{L}^{2^{m-1}} + \partial_t)K^{(m-1)} = \delta$ and that $K^{(m-1)}(\cdot,t) \in \mathcal{S}$. But now we can repeat this reasoning with m replaced successively by $m-1,\ldots,2$, concluding at last that $K(\cdot,t) \in \mathcal{S}$.

Now let $\phi(x) = K(x,1)$. By Lemma 4.22(c), we have $\phi_t(x) = K(x,t^D)$. On the other hand, since $A_tA_s = A_{t+s}$ it follows that $K(\cdot,t+s) = K(\cdot,t)*K(\cdot,s)$ and hence

$$\phi_t*\phi_s = K(\cdot,t^D+s^D) = \phi_s*\phi_t.$$

Also, if $f \in L^2$,

$$\|f*\phi_t - f\|_2 = \|A_{t^D}f - f\|_2 \to 0 \quad \text{as} \quad t \to 0,$$

which implies that $\int\phi = 1$ by Proposition 1.20. Thus ϕ is a commutative approximate identity. #

Finally, we present another construction of commutative approximate identities on groups which have certain symmetry properties.

(4.27) PROPOSITION. (a) _Let G be a unimodular locally compact group which possesses a measure-preserving antiautomorphism τ (i.e. $\tau(xy) = \tau(y)\tau(x)$) and a family Σ of measure-preserving automorphisms such that for each $x \in G$ there exists $\sigma_x \in \Sigma$ such that $\tau(x) = \sigma_x(x)$. If $\phi,\psi \in L^1(G)$ are invariant under τ and under every $\sigma \in \Sigma$ then $\phi*\psi = \psi*\phi$._

(b) _Suppose in addition that G is a homogeneous group and that τ and the elements of Σ commute with dilations. Then every $\phi \in L^1(G)$ which is invariant under τ and under every $\sigma \in \Sigma$ satisfies $\phi_s*\phi_t = \phi_t*\phi_s$ for all $s,t > 0$._

(<u>Remark</u>: Of course, unless Σ is reasonably small there may not exist any such ϕ's.)

<u>Proof</u>: It suffices to prove (a), as (b) is an immediate consequence thereof. It is easily verified that for any $\phi,\psi \in L^1(G)$,

$$(\phi \circ \tau)*(\psi \circ \tau) = (\psi*\phi) \circ \tau,$$

$$(\phi \circ \sigma)*(\psi \circ \sigma) = (\phi*\psi) \circ \sigma \qquad \text{for } \sigma \in \Sigma.$$

Hence if ϕ and ψ are invariant under τ and under every $\sigma \in \Sigma$, for almost every $x \in G$ (namely those x for which the convolution integral converges), we have

$$\phi*\psi(x) = (\phi \circ \tau)*(\psi \circ \tau)(x) = (\psi*\phi)(\tau(x)) = (\psi*\phi)(\sigma_x(x))$$

$$= (\psi \circ \sigma_x)*(\phi \circ \sigma_x)(x) = \psi*\phi(x). \quad \#$$

The example we have in mind is the Heisenberg group H_n defined in Chapter 1, Section A, with the standard dilations

$$\delta_r(z_1,\ldots,z_n,t) = (rz_1,\ldots,rz_n,r^2t).$$

We define $\tau : H_n \to H_n$ by

$$\tau(z_1,\ldots,z_n,t) = (\bar{z}_1,\ldots,\bar{z}_n,t),$$

and for each α in the n-torus $T^n = \{\alpha \in \mathbb{C}^n : |\alpha_j| = 1 \text{ for all } j\}$, we define $\sigma_\alpha : H_n \to H_n$ by

$$\sigma_\alpha(z_1,\ldots,z_n,t) = (\alpha_1 z_1,\ldots,\alpha_n z_n,t).$$

It is then easily verified that τ and $\Sigma = \{\sigma_\alpha : \alpha \in T^n\}$ satisfy all the conditions of Proposition 4.27. The functions which are invariant under τ and under every $\sigma \in \Sigma$ are precisely the <u>polyradial</u> functions on H_n, that is, the functions ϕ for which there exists a function ϕ_0 on $[0,\infty)^n \times \mathbb{R}$ such that

$$\phi(z_1,\ldots,z_n,t) = \phi_0(|z_1|,\ldots,|z_n|,t).$$

We therefore conclude:

(4.28) PROPOSITION. *If $\phi \in \mathcal{S}(H_n)$ is polyradial and $\int \phi = 1$ then ϕ is a commutative approximate identity.*

Notes and References

For $G = \mathbb{R}^n$, the results on the equivalence of various maximal functions are due to Fefferman and Stein [1], whose arguments we have followed in proving Theorems 4.10 and 4.12. For $G = \mathbb{R}^n$ with nonisotropic dilations, these results are (implicitly or explicitly) in Calderón and Torchinsky [1]. For polyradial functions on the Heisenberg group, they are due to Geller [1], [2], who proved Proposition 4.4 for such functions by using the group Fourier transform. Our use of commutative approximate identities and in particular Theorem 4.9 is novel, but it should be noted that such approximations have previously been found useful in other situations: see, for example, Coifman and Weiss [1], section III.3.

Recently, Uchiyama [2] has found a new proof of the theorem "$M_\phi^0 f \in L^p$ implies $M_{(N)} f \in L^p$" for $\phi \in \mathcal{S}$, $\int \phi = 1$ on $\mathbb{R}^n$ with nonisotropic dilations, in which the convolution integral $f*\phi_t(x)$ can even be replaced by more general integrals of the form $\int f(y)\Phi(x,y,t)dy$ where Φ satisfies certain conditions. Uchiyama [1] has also proved a version of this result on general spaces of homogeneous type, for p very close to 1.

The results of this chapter lead to two questions: First, whether the analogues of Theorems 4.10 - 4.12 hold for more general approximate identities. Secondly, whether commutative approximate identities exist on any homogeneous group. As for the construction of commutative approximate identities via homogeneous hypoelliptic differential operators, this works only for graded groups, since such operators exist only for graded groups. (This fact was proved by Miller [1].)

A result related to Proposition 4.27 may be found in Kaplan and Putz [1].

CHAPTER 5

Duals of H^p spaces: Campanato Spaces

Campanato spaces are function spaces defined in terms of approximation by polynomials on balls, generalizing the idea of bounded mean oscillation introduced by John and Nirenberg. Among them are the duals of the H^p spaces for $0 < p \leq 1$, and they are also of considerable interest in their own right. In this chapter we prove the duality theorem for H^p and then investigate the relationships between the Campanato spaces and the more familiar Lipschitz classes.

A. The Dual of H^p

In this section we compute the dual space of H^p, $0 < p \leq 1$. Our description of $(H^p)^*$ will be of the following nature. If (p,q,a) is admissible,[†] the finite linear combinations of (p,q,a)-atoms are dense in H^p, so an element of $(H^p)^*$ is completely determined by its action on such functions. We shall describe this action explicitly, obtaining for each q,a a different characterization of $(H^p)^*$. The fact that these characterizations are equivalent will then lead to interesting results.

If L is a linear functional on H^p and (p,q,a) is admissible, we define

$$\mu^p_{q,a}(L) = \sup\{|Lf| : f \text{ is a } (p,q,a)\text{-atom}\}.$$

[†]Recall that this requires that a is sufficiently large, and $q > p$; the precise conditions are on p. 71.

(5.1) LEMMA. $\mu^p_{q,a}(L) = \sup\{|Lf| : \rho^p_{q,a}(f) \le 1\}$.

Proof: By (2.17), for each $\varepsilon > 0$ there is a (p,q,a)-atom f with $\rho^p_{q,a}(f) > 1-\varepsilon$, so $\mu^p_{q,a}(L) \ge \sup\{|Lf| : \rho^p_{q,a}(f) \le 1\}$. On the other hand, if $f \in H^p$ and $\rho^p_{q,a}(f) \le 1$, for each $\varepsilon > 0$ there is an atomic decomposition $f = \Sigma\lambda_i f_i$ with $\Sigma\lambda^p_i < 1+\varepsilon$, so

$$|Lf| \le \Sigma\lambda_i |Lf_i| \le \mu^p_{q,a}(L)(\Sigma\lambda^p_i)^{1/p} \le \mu^p_{q,a}(L)(1+\varepsilon)^{1/p},$$

which shows that $\sup\{|Lf| : \rho^p_{q,a}(f) \le 1\} \le \mu^p_{q,a}(L)$. #

The usual elementary arguments show that L is continuous on H^p if and only if $\mu^p_{q,a}(L) < \infty$, and that $\mu^p_{q,a}$ is a norm on $(H^p)^*$ which makes $(H^p)^*$ into a Banach space. We remark that the proof of Lemma 5.1 shows that every $L \in (H^p)^*$ extends continuously to the Banach space obtained by completin H^p with respect to the norm

$$\|f\| = \inf\{\Sigma\lambda_i : f = \Sigma\lambda_i f_i \text{ is an atomic decomposition of } f\}.$$

Thus when $p < 1$ we lose information in passing from H^p to $(H^p)^*$.

We now define the Campanato spaces $C^\alpha_{q,a}$. Let $\mathcal{B}$ denote the collection of all open balls in G. If $\alpha \ge 0$, $1 \le q \le \infty$, and $a \in \Delta$, we define $C^\alpha_{q,a}$ to be the space of all locally L^q functions u on G such that

$$(5.2) \qquad \nu^\alpha_{q,a}(u) \equiv \sup_{B\in\mathcal{B}} \{\inf_{P\in\mathcal{P}_a} |B|^{-\alpha/Q}[|B|^{-1} \int_B |u(x) - P(x)|^q dx]^{1/q}\} < \infty.$$

(If $q = \infty$, of course,

$$\nu^\alpha_{\infty,a}(u) \equiv \sup_{B\in\mathcal{B}} \{\inf_{P\in\mathcal{P}_a} \operatorname{ess\,sup}_{x\in B} |B|^{-\alpha/Q} |u(x) - P(x)|\}.)$$

We identify two elements of $C^{\alpha}_{q,a}$ if they are equal almost everywhere. The following facts may be readily verified by the reader: (i) $\nu^{\alpha}_{q,a}$ is a seminorm on $C^{\alpha}_{q,a}$. (ii) $\mathcal{P}_a \subset C^{\alpha}_{q,a}$, and $\nu^{\alpha}_{q,a}(u) = 0$ if and only if $u \in \mathcal{P}_a$. Thus, $C^{\alpha}_{q,a}/\mathcal{P}_a$ is a normed linear space.

We shall investigate the properties of $C^{\alpha}_{q,a}$ more thoroughly in Sections B and C. For the moment, we note that if we restrict the balls B in (5.2) to satisfy $|B| \geq 1$, then (5.2) is a condition on the behavior of u at infinity which is surely satisfied if u is bounded (then one can take $P = 0$). On the other hand, if we restrict the balls to satisfy $|B| \leq 1$, (5.2) is a smoothness condition on u which is satisfied if u has sufficiently many bounded continuous derivatives (then one can take P to be a Taylor polynomial of u based at the center of B).

Our duality theorem is the following.

(5.3) THEOREM. *If* (p,q,a) *is admissible, then*

$$(H^p)^* = (H^p_{q,a})^* \cong C^{\alpha}_{q',a}/\mathcal{P}_a, \quad \text{where} \quad q^{-1} + (q')^{-1} = 1 \quad \text{and} \quad \alpha = Q(p^{-1}-1).$$

More precisely, if $u \in C^{\alpha}_{q',a}$ *and* f *is a finite linear combination of* (p,q,a)-*atoms, let* $L_u f = \int uf$. *Then* L_u *extends continuously to* H^p, *and every* $L \in (H^p)^*$ *is of this form. Moreover,* $\nu^{\alpha}_{q',a}(u) = \mu^p_{q,a}(L_u)$ *for every* $u \in C^{\alpha}_{q',a}$ (*and, in particular,* $L_u = 0$ *if and only if* $u \in \mathcal{P}_a$).

Proof: Suppose $u \in C^{\alpha}_{q',a}$ and suppose f is a (p,q,a)-atom associated to the ball B. Then $\int uf = \int (u-P)f$ for all $P \in \mathcal{P}_a$, so

$$\left|\int uf\right| = \inf_{P\in\mathcal{P}_a} \left|\int (u-P)f\right| \le \left[\int_B |f|^q\right]^{1/q} \left[\inf_{P\in\mathcal{P}_a} \int_B |u-P|^{q'}\right]^{1/q'}$$

$$\le |B|^{(1/q)-(1/p)}\ |B|^{(\alpha/Q)+(1/q')}\ \nu^{\alpha}_{q',a}(u) = \nu^{\alpha}_{q',a}(u).$$

Thus $\mu^p_{q,a}(L_u) \le \nu^{\alpha}_{q',a}(u)$.

To prove the converse, for each $B\in\mathcal{B}$ let $\pi_B : L^1(B) \to \mathcal{P}_a$ be the natural projection characterized by the condition

$$\int_B (\pi_B f)P = \int_B fP \quad \text{for all } f\in L^1(B),\ P\in\mathcal{P}_a.$$

Also, let $L^q_0(B) = \{f\in L^q(B) : \pi_B f = 0\}$. We shall identify $L^q(B)$ with the subspace of $L^q = L^q(G)$ consisting of functions which vanish outside B. With this identification, we remark that if $f\in L^q_0(B)$ then $|B|^{(1/q)-(1/p)}\|f\|^{-1}_q$ is a (p,q,a)-atom.

Suppose $L\in (H^p)^*$. By the preceding remark, if $f\in L^q_0(B)$ then

$$|Lf| \le \mu^p_{q,a}(L)|B|^{(1/p)-(1/q)}\|f\|_q. \tag{5.4}$$

Hence L is a bounded linear functional on $L^q_0(B)$, which can be extended by the Hahn-Banach theorem to an element of $(L^q(B))^*$. Thus if $q < \infty$, there exists $v\in L^{q'}(B)$ such that $Lf = \int fv$ for all $f\in L^q_0(B)$. This is true even if $q = \infty$, since $H^p = H^p_{r,a}$ for $1 < r < \infty$, so any $L\in (H^p)^*$ must define an element of $(L^r_0(B))^*$. If v' is another element of $L^{q'}(B)$ such that $Lf = \int fv'$ for all $f\in L^q_0(B)$, then for all $g\in L^q(B)$,

$$0 = \int_B (g-\pi_B g)(v-v') = \int_B g(v-v') - \int_B (\pi_B g)(\pi_B(v-v'))$$

$$= \int_B g(v-v') - \int_B g(\pi_B(v-v')) = \int_B g[(v-v') - \pi_B(v-v')],$$

which implies that $v-v' = \pi_B(v-v') \in P_a$.

Now, for $k = 1,2,3,\ldots$ let u_k be the element of $L^{q'}(B(k,0))$ such that $Lf = \int fu_k$ for all $f \in L^q_0(B(k,0))$ and $\pi_{B(1,0)}u_k = 0$. The preceding arguments show that if $j < k$ then $u_k|B(j,0) = u_j$. We can therefore define a locally $L^{q'}$ function u on G by setting $u(x) = u_k(x)$ for $u \in B(k,0)$, and it follows that $Lf = \int fu$ for all finite linear combinations f of (p,q,a)-atoms. Moreover, by (5.4), for any $B \in \mathcal{B}$ the norm of u as a linear functional on $L^q_0(B)$ satisfies

$$\|u\|_{L^q_0(B)^*} \leq |B|^{(1/p)-(1/q)} \mu^p_{q,a}(L).$$

But by elementary functional analysis,

$$\|u\|_{L^q_0(B)^*} = \inf_{P \in P_a} [\int_B |u-P|^{q'}]^{1/q'}.$$

Hence, for $\alpha = Q(p^{-1}-1)$,

$$\nu^\alpha_{q',a}(u) \leq \sup_{B \in \mathcal{B}} |B|^{-(\alpha/Q)-(1/q')} |B|^{(1/p)-(1/q)} \mu^p_{q,a}(L)$$

$$= \mu^p_{q,a}(L).$$

Therefore $u \in C^\alpha_{q',a}$, and we are done. #

(5.5) COROLLARY. (a) _If_ $\alpha \geq 0$, $1 \leq q < \infty$, $1 \leq r \leq \infty$, $a \in \Delta$, _and_ $a \geq \max\{b \in \Delta : b \leq \alpha\}$, _then_ $C^\alpha_{q,a} = C^\alpha_{r,a}$ _provided that either_ $r < \infty$ _or_ $\alpha > 0$.

(b) _If_ $\alpha \geq 0$, $1 \leq q \leq \infty$ (_with_ $q < \infty$ _if_ $\alpha = 0$), $a_1, a_2 \in \Delta$, _and_ $a_1 > a_2 \geq \max\{b \in \Delta : b \leq \alpha\}$, _then for any_ $u \in C^\alpha_{q,a_1}$ _there exists_ $P \in P_{a_1}$ _such that_ $u-P \in C^\alpha_{q,a_2}$.

Proof: The hypotheses guarantee that (p,q',a), (p,r',a), (p,q',a_1), and (p,q',a_2) are all admissible, where q' and r' are conjugate to q and r and $p = Q/(Q+\alpha)$. The assertions then follow easily from Theorem 5.3. #

Remark: One should not try to dualize the interpolation theorem (3.34) to obtain interpolation results between $C^{\alpha}_{q,a}$ and L^p for $\alpha > 0$, because H^p is not a Banach space for $p < 1$. In fact, in view of the identification of Campanato spaces and Lipschitz classes which we shall prove below, such results are known to be false: cf. Stein and Zygmund [1].

B. BMO

We now examine more closely the nature of the spaces $C^{\alpha}_{q,a}$, beginning with the case $\alpha = a = 0$. It is an easy exercise to show that $C^0_{\infty,0} = L^{\infty}$. For $q < \infty$, however, $C^0_{q,0}$ is larger than L^{∞}.

If u is a locally integrable function on G and B is a ball, we set

$$m_B u = |B|^{-1} \int_B u(x)dx.$$

We then define BMO ("bounded mean oscillation") to be the space of all locally integrable functions u on G such that

$$\|u\|_{BMO} \equiv \sup_{B\in\mathcal{B}} |B|^{-1} \int_B |u(x) - m_B u|dx < \infty.$$

(5.6) PROPOSITION. BMO $= C^0_{1,0}$, and $\nu^0_{1,0} \le \|\ \|_{BMO} \le 2\nu^0_{1,0}$.

Proof: Obviously $\nu^0_{1,0}(u) \leq \|u\|_{BMO}$, so $BMO \subset C^0_{1,0}$. Conversely, suppose $u \in C^0_{1,0}$ and let $A = \nu^0_{1,0}(u)$. Then for each $\varepsilon > 0$ and each $B \in \mathcal{B}$ there is a constant c_B such that

$$|B|^{-1} \int_B |u(x) - c_B|\,dx \leq A+\varepsilon.$$

Hence,

$$\begin{aligned} |B|^{-1} \int_B |u(x) - m_B u|\,dx &\leq |B|^{-1} \int_B |u(x) - c_B|\,dx + |B|^{-1} \int_B |c_B - m_B u|\,dx \\ &\leq A+\varepsilon + |c_B - m_B u| \\ &\leq A+\varepsilon + |B|^{-1} \left|\int_B (c_B - u(x))\,dx\right| \\ &\leq 2(A+\varepsilon), \end{aligned}$$

so that $u \in BMO$ and $\|u\|_{BMO} \leq 2\nu^0_{1,0}(u)$. #

(5.7) COROLLARY. $(H^1)^* = BMO/\mathcal{P}_0$.

(5.8) COROLLARY. _If_ $1 \leq a < \infty$ _and_ $a \in \Delta$ _then_ $C^0_{q,a} = BMO + \mathcal{P}_a$. _In particular,_ $C^0_{q,0} = BMO$, _so elements of_ BMO _are locally in_ L^q _for all_ $q < \infty$.

Proof: Apply Corollary 5.5.

Remark: The assertion that $BMO \subset L^q_{loc}$ can be improved: see Corollary 5.16 below.

We obtained the duality theorem (5.7) by using the atomic decomposition theorem for H^1. It is of interest that these theorems are equivalent, in the following sense. The proof of Theorem 5.3 (together with Proposition 5.6) shows, in any event, that $(H^1_{\infty,0})^* = BMO/P_0$, while by Proposition 2.15 we have $H^1_{\infty,0} \subset H^1$. Both of these arguments are relatively elementary. Thus if we take the equation $(H^1)^* = BMO/P_0$ as given, we have $H^1_{\infty,0} \subset H^1$ and $(H^1_{\infty,0})^* = (H^1)^*$. Since $H^1_{\infty,0}$ and H^1 are Banach spaces, the Hahn-Banach theorem implies that $H^1_{\infty,0}$ is dense in H^1, while the uniform boundedness principle implies that subsets of $H^1_{\infty,0}$ which are bounded in H^1 are bounded in $H^1_{\infty,0}$, so that the norms are equivalent. Thus $H^1 = H^1_{\infty,0}$.

For future reference we note the following property of BMO functions:

(5.9) PROPOSITION. _if_ $u \in BMO$ _then for any_ $\varepsilon > 0$,

$$\int |u(x)|(1+|x|)^{-Q-\varepsilon}dx < \infty.$$

Proof: Let $A = \|u\|_{BMO}$, and for $k \in \mathbb{N}$ let

$$\mu_k = m_{B(2^k,0)} u = |B(2^k,0)|^{-1} \int_{B(2^k,0)} u(x)dx.$$

Thus

$$\int_{B(2^k,0)} |u(x) - \mu_k| dx \le A2^{kQ}.$$

We have

$$|\mu_{k-1} - \mu_k| = |B(2^{k-1},0)|^{-1} \left|\int_{B(2^{k-1},0)} (u(x) - \mu_k)dx\right|$$

$$\le 2^{(1-k)Q}A2^{kQ} = A2^Q,$$

and hence $|\mu_k - \mu_0| \leq A2^Q k$, so that for $k \geq 1$

$$\int_{B(2^k,0)} |u(x) - \mu_0|\,dx \leq A2^{kQ} + A2^Q k\cdot 2^{kQ} = A(1+2^Q k)2^{kQ}.$$

Therefore,

$$\int |u(x) - \mu_0|(1+|x|)^{-Q-\varepsilon}dx$$

$$\leq \int_{|x|\leq 1} |u(x) - \mu_0|\,dx + \sum_{k=1}^{\infty} \int_{2^{k-1}\leq|x|\leq 2^k} |u(x) - \mu_0| 2^{-(Q+\varepsilon)(k-1)}dx$$

$$\leq A + \sum_{k=1}^{\infty} A(1+2^Q k)2^{kQ}2^{-(Q+\varepsilon)(k-1)}$$

$$= A(1+2^{Q+\varepsilon} \sum_{k=1}^{\infty} (1+2^Q k)2^{-\varepsilon k}) < \infty,$$

which yields the desired result since $\int (1+|x|)^{-Q-\varepsilon}dx < \infty$. #

We can obtain more insight into BMO by introducing the <u>sharp function</u>. If u is locally integrable on G, we set

$$u^{\#}(x) = \sup\{|B|^{-1}\int_B |u(y) - m_B u|\,dy : B \text{ is a ball containing } x\}.$$

Clearly we have

$$\|u\|_{BMO} = \|u^{\#}\|_{\infty}.$$

The following result expresses the duality of H^1 and BMO in terms of maximal functions.

(5.10) THEOREM. There is a constant C such that for all $f \in H^1$ and $u \in BMO$,

$$\left|\int f(x)u(x)dx\right| \le C \int Mf(x)u^{\#}(x)dx \qquad (M = M_{(1)}).$$

Proof: If $f \in H^1$, the proof of Theorem 3.28 yields an atomic decomposition $f = \Sigma_{ik}\lambda_i^k a_i^k$ with the following properties:

(i) a_i^k is a $(1,\infty,0)$-atom associated to a ball $B_i^k (= B(T_2 r_i^k, x_i^k)$ in the notation of Chapter 3).

(ii) $\lambda_i^k = C2^k|B_i^k|$ where C is an absolute constant.

(iii) For each k, $\bigcup_i B_i^k = \Omega^k$ where $\Omega^k = \{x : Mf(x) > 2^k\}$, and there is an absolute contant L such that each $x \in \Omega^k$ is contained in at most L of the balls B_i^k.

If $u \in BMO$, then, by (i) we have

$$\begin{aligned}\left|\int f(y)u(y)dy\right| &= \left|\Sigma_{ik}\lambda_i^k \int a_i^k(y)u(y)dy\right| \\ &= \left|\Sigma_{ik}\lambda_i^k \int a_i^k(y)(u(y) - m_{B_i^k}u)dy\right| \\ &\le \Sigma_{ik}\lambda_i^k|B_i^k|^{-1}\int_{B_i^k}|u(y) - m_{B_i^k}u|dy \\ &\le \Sigma_{ik}\lambda_i^k u^{\#}(x) \qquad \text{for } x \in B_i^k,\end{aligned}$$

so that, in view of (ii) and (iii),

$$\left|\int f(y)u(y)dy\right| \le \Sigma_{ik}\lambda_i^k|B_i^k|^{-1}\int_{B_i^k} u^{\#}(x)dx$$

$$\le C\Sigma_{ik}2^k\int_{B_i^k} u^{\#}(x)dx$$

$$\le LC\Sigma_k 2^k\int_{\Omega^k} u^{\#}(x)dx.$$

Since $\Omega^{k+1} \subset \Omega^k$ for all k, we have

$$\Sigma_k 2^k \chi_{\Omega^k} = \Sigma_k 2^{k+1}\chi_{\Omega^k\setminus\Omega^{k+1}} \le 2Mf,$$

and therefore

$$\left|\int f(y)u(y)dy\right| \le LC\int(\Sigma_k 2^k\chi_{\Omega^k}(x))u^{\#}(x)dx$$

$$\le 2LC\int Mf(x)u^{\#}(x)dx. \quad \#$$

We next prove a localized version of this result, from which we shall deduce the John-Nirenberg inequality for BMO functions.

(5.11) LEMMA. *Suppose* $f \in H^1$ *and* f *is supported in* $B = B(r_0,x_0)$. *There exist* $A \ge 2\gamma$ *and* $C > 0$, *independent of* f, r_0, *and* x_0, *such that if* $\tilde{B} = B(Ar_0,x_0)$,

$$\int_{\tilde{B}} Mf(x)dx \ge C\rho^1_{\infty,0}(f) \qquad (M = M_{(1)}).$$

Proof: Without loss of generality we may assume that $x_0 = 0$. Suppose $\phi \in S$ and $\|\phi\|_{(1)} \le 1$. By theorem 1.50 (with $a = 0$), if $|x| \ge 2\gamma r_0$ we have

$$|f*\phi_t(x)| = \left|\int f(y)[\phi_t(y^{-1}x) - \phi_t(x)]dy\right|$$

$$\leq C_1 \int_B |f(y)||y||x|^{-Q-1}dy$$

$$\leq C_1 r_0 |x|^{-Q-1} \int |f(y)|dy$$

$$\leq C_1 r_0 \rho^1_{\infty,0}(f)|x|^{-Q-1}.$$

Therefore, if $A \geq 2\gamma$,

$$\int_{|x|>Ar_0} Mf(x)dx \leq C_2 r_0 \rho^1_{\infty,0}(f) \int_{Ar_0}^{\infty} r^{-2}dr$$

$$= (C_2/A)\rho^1_{\infty,0}(f).$$

On the other hand, by Theorem 3.28 there exists C_3 such that

$$\int_G Mf(x)dx \geq C_3 \rho^1_{\infty,0}(f).$$

Thus if we take $A = \max(2C_2/C_3, 2\gamma)$ we have

$$\int_{\tilde{B}} Mf(x)dx \geq (C_3/2)\rho^1_{\infty,0}(f). \quad \#$$

(5.12) THEOREM. Let $A, B, \tilde{B}$ be as in Lemma 5.11. There exists $C > 0$ such that for all $f \in H^1$ supported in B and all $u \in BMO$,

$$\left|\int f(x)u(x)dx\right| \leq C \int_{\tilde{B}} Mf(x)u^{\#}(x)dx \qquad (M = M_{(1)}).$$

<u>Proof</u>: Let $f = \Sigma\lambda_i^k a_i^k$ be as in the proof of Theorem 5.10. In addition to properties (i) - (iii) of this decomposition listed above, we need one more:

(iv) If B_i^k does not intersect the support of f then $a_i^k = 0$.

Let

$$I_1 = \{(i,k) : B_i^k \cap B \neq \emptyset \text{ and radius } (B_i^k) < r_0\}$$

$$I_2 = \{(i,k) : B_i^k \cap B \neq \emptyset \text{ and radius } (B_i^k) \geq r_0\}.$$

Then in view of (iv), we have

$$f = f_1 + f_2, \quad \text{where} \quad f_j = \Sigma_{I_j} \lambda_i^k a_i^k.$$

On the one hand, since $A \geq 2\gamma$ we see that f_1 is supported in $\tilde{B}$. The proof of Theorem 5.10, with Ω^k redefined to be $\{x \in \tilde{B} : Mf(x) > 2^k\}$, then shows that

$$(5.13) \qquad \left|\int f_1(x)u(x)dx\right| \leq C_1 \int_{\tilde{B}} Mf(x)dx.$$

On the other hand, if $(i,k) \in I_2$ we have $\|a_i^k\|_\infty \leq |B_i^k|^{-1} \leq r_0^{-Q}$, and hence

$$\|f_2\|_\infty \leq r_0^{-Q} \Sigma_{I_2} \lambda_i^k \leq C_2 r_0^{-Q} \rho^1(f) \leq C_3 r_0^{-Q} \rho_{\infty,0}^1(f).$$

Moreover, $\int f_2 = 0$ and $f_2 = f - f_1$ is supported in $\tilde{B}$. Thus

$$\begin{aligned}\left|\int f_2(y)u(y)dy\right| &= \left|\int f_2(y)(u(y) - m_{\tilde{B}}u)dy\right| \\ &\leq C_3 r_0^{-Q} \rho_{\infty,0}^1(f) \int_{\tilde{B}} |u(y) - m_{\tilde{B}}u|\, dy \\ &\leq C_3 A^Q \rho_{\infty,0}^1(f) u^\#(x) \quad \text{for} \quad x \in \tilde{B}.\end{aligned}$$

If we multiply both sides of this inequality by

$$Mf(x) \Big/ \int_{\tilde{B}} Mf(y)dy,$$

integrate over $\tilde{B}$, and apply Lemma 5.11, we obtain

$$(5.14) \qquad \left|\int f_2(y)u(y)dy\right| \leq C_3 A^Q \rho^1_{\infty,0}(f)\left[\int_{\tilde{B}} Mf(y)dy\right]^{-1} \int_{\tilde{B}} Mf(x)u^{\#}(x)dx$$

$$\leq C_4 \int_{\tilde{B}} Mf(x)u^{\#}(x)dx.$$

Combining (5.13) and (5.14), we are done. #

(5.15) THEOREM. <u>There exist constants</u> C, C' <u>such that for every</u> $u \in BMO$, <u>every ball</u> B, <u>and every</u> $\alpha > 0$,

$$|\{x \in B : |u(x) - m_B u| > \alpha\}| \leq C|B|\exp(-C'\alpha/\|u\|_{BMO}).$$

<u>Proof</u>: It clearly suffices to assume that $\|u\|_{BMO} = 1$ and that u is real-valued. Given a ball B and $\alpha > 0$, let

$$E = \{x \in B : u(x) - m_B u > \alpha\}.$$

Then

$$\int_B \chi_E(x)(u(x) - m_B u)dx = \int_B (\chi_E(x) - m_B \chi_E)u(x)dx.$$

Now, the function

$$f(x) = \chi_E(x) - (m_B\chi_E)\chi_B(x) = \chi_E(x) - |B|^{-1}|E|\chi_B(x)$$

is bounded and supported in B, and $\int f = 0$. Thus f is a constant multiple of a $(1,\infty,0)$-atom, so by Theorem 5.12, if $\tilde{B}$ is the ball concentric with $\tilde{B}$ whose radius is A times as large as that of B,

$$\left|\int_B \chi_E(x)(u(x) - m_B u)dx\right| \le C_1 \int_{\tilde{B}} Mf(x)dx \qquad (\text{since } \|u^\#\|_\infty = 1).$$

Also,

$$Mf(x) \le M\chi_E(x) + |B|^{-1}|E|M\chi_B(x) \le M\chi_E(x) + C_2|B|^{-1}|E|,$$

so that

$$\int_{\tilde{B}} Mf(x)dx \le \int_{\tilde{B}} M\chi_E(x)dx + C_2 A^Q|E| \le C_3 \int_{\tilde{B}} M\chi_E(x)dx,$$

and hence

$$\left|\int_B \chi_E(x)(u(x) - m_B u)dx\right| \le C_4 \int_{\tilde{B}} M\chi_E(x)dx.$$

Therefore, if $1 < p < \infty$ and q is the conjugate exponent to p, by the definition of E, Hölder's inequality, and the maximal theorem (2.4),

$$\alpha|E| \le \left|\int_B \chi_E(x)(u(x) - m_B u)dx\right| \le C_4\left(\int_{\tilde{B}} M\chi_E(x)^q dx\right)^{1/q} |\tilde{B}|^{1/p}$$

$$\le C_5 p\|\chi_E\|_q |B|^{1/p} = C_5 p|E|^{1/q}|B|^{1/p}.$$

In other words,

$$|E| \le (C_5 p/\alpha)^p |B|.$$

Now, if $\alpha \le 2C_5$ the desired inequality is trivial. If $\alpha > 2C_5$, let $p = \alpha/2C_5$. Then

$$|E| \le (1/2)^p|B| = e^{-C'\alpha}|B| \quad \text{where } C' = (\log 2)/2C_5.$$

The same argument yields the same estimate for the measure of

$$F = \{x \in B : u(x) - m_B u < -\alpha\},$$

so we are done. #

(5.16) COROLLARY. *If* $u \in BMO$ *and* $\varepsilon < C'$ *then* $e^{\varepsilon|u|}$ *is locally integrable on* G.

C. Lipschitz Classes

If $\alpha > 0$, the elements of $C^\alpha_{q,a}$ are continuous (after correction on a set of measure zero) and in fact belong to certain Lipschitz classes, depending mainly on the size of α. In the case where G is a stratified group, we obtain below a precise global characterization of $C^\alpha_{q,a}$ as a Lipschitz space (although our results for $\alpha \in \mathbb{N}$ are not quite complete). For the general case, we content ourselves with stating the following description of the local smoothness properties of the elements of $C^\alpha_{q,a}$, which follows from results of Krantz [1], [2].

Notation: If $\alpha > 0$, we denote by $\Lambda_\alpha(-1,1)$ the classical Lipschitz space of order α on the interval $(-1,1)$ (cf. Stein [2]). If $x \in G$ and $1 \leq j \leq n$, we define $\gamma^j_x : (-1,1) \to G$ by $\gamma^j_x(t) = x \cdot \exp(tX_j)$.

(5.17) PROPOSITION. *Suppose* $\alpha > 0$, $1 \leq q \leq \infty$, *and* $a \in \Delta$. *If* $u \in C^\alpha_{q,a}$ *and* $d(I) < \alpha$ *then* $X^I u$ *is continuous, and moreover*

$$(X^I u) \circ \gamma^j_x \in \Lambda_{(\alpha - d(I))/d_j}(-1,1) \quad \textit{for} \quad x \in G,\ 1 \leq j \leq n.$$

For the remainder of this section we assume that G is a stratified group. We shall work from scratch, without using Proposition 5.17. First, a few details to set the stage.

(1) Since $\Delta = \mathbb{N}$ for a stratified group, we shall denote the elements of Δ by N rather than a.

(2) By Corollary 5.5, we have $C^{\alpha}_{q,N} = C^{\alpha}_{\infty,N}$ for $\alpha > 0$ and $N \geq [\alpha]$. (Actually, $C^{\alpha}_{q,N} = C^{\alpha}_{\infty,N}$ for $\alpha > 0$ no matter what N is, as can be shown by a modification of the arguments given below. See the notes at the end of the chapter.) We shall therefore assume that $q = \infty$ and suppress the subscript q henceforth. Thus,

$$C^{\alpha}_{N} = C^{\alpha}_{\infty,N}, \qquad \nu^{\alpha}_{N} = \nu^{\alpha}_{\infty,N}.$$

(3) Also by Corollary 5.5, if $N_1 \geq N_2 \geq [\alpha]$ we have $C^{\alpha}_{N_1} = C^{\alpha}_{N_2} + P_{N_1}$, so it suffices to consider $N \leq [\alpha]$. (The cases $N < [\alpha]$ are not relevant to H^p theory, but we shall obtain results for them which are of interest in their own right.)

(4) In the definition of the seminorm ν^{α}_{N} it will henceforth be convenient to identify the balls in G by their center and radius. Thus,

$$\nu^{\alpha}_{N}(u) = \sup_{r>0,x\in G} \inf_{P\in P_N} \text{ess sup}_{y\in B(r,x)} r^{-\alpha}|u(x) - P(x)|.$$

Next, we define the Lipschitz classes with which we shall be dealing. These are homogeneous versions of the spaces called Γ_{α} in Folland [1], [3]; accordingly, we shall denote them by Γ^{hom}_{α}. We recall from Chapter 1,

Section C that C^k is the set of continuous functions u on G such that $X^I u$ is continuous for $d(I) \le k$. If $\alpha > 0$ and α is not an integer, we define

$$\Gamma_\alpha^{hom} = \{u \in C^{[\alpha]} : |u|_\alpha < \infty\}, \quad \text{where}$$

$$|u|_\alpha = \sup_{d(I)=[\alpha]} \sup_{x,y \in G} |X^I u(xy) - X^I u(x)|/|y|^{\alpha-[\alpha]}.$$

If α is a positive integer, there are two reasonable definitions of a Lipschitz class of order α. The one which occurs most frequently is the "Zygmund class"

$$\Gamma_\alpha^{hom} = \{u \in C^{\alpha-1} : |u|_\alpha < \infty\}, \quad \text{where}$$

$$|u|_\alpha = \sup_{d(I)=\alpha-1} \sup_{x,y \in G} |X^I u(xy) + X^I u(xy^{-1}) - 2X^I u(x)|/|y|.$$

However, we shall also encounter the "naive" Lipschitz class

$$\hat{\Gamma}_\alpha^{hom} = \{u \in C^{\alpha-1} : |u|_\alpha^\wedge < \infty\}, \quad \text{where}$$

$$|u|_\alpha^\wedge = \sup_{d(I)=\alpha-1} \sup_{s,y \in G} |X^I u(xy) - X^I u(x)|/|y|.$$

For the sake of completeness, we mention that for $\alpha > 0$, the space Γ_α of Folland [1], [3] is the set of all $u \in \Gamma_\alpha^{hom}$ such that $X^I u$ is bounded for $0 \le d(I) < \alpha$.

The functionals $u \to |u|_\alpha$, $u \to |u|_\alpha^\wedge$ are obviously seminorms on Γ_α^{hom}, $\hat{\Gamma}_\alpha^{hom}$. We define

$$N_\alpha = \{u \in \Gamma_\alpha^{hom} : |u|_\alpha = 0\}, \qquad \hat{N}_\alpha = \{u \in \hat{\Gamma}_\alpha^{hom} : |u|_\alpha^\wedge = 0\}.$$

(5.18) PROPOSITION. *If* α *is not an integer*, $N_\alpha = P_{[\alpha]}$. *If* α *is an integer*, $\hat{N}_\alpha = P_{\alpha-1}$.

Proof: If $u \in P_k$ then $X^I u \in P_0$ for $d(I) = k$. Conversely, if $u \in C^k$ and $X^I u \in P_0$ for $d(I) = k$ then $X^J u = 0$ for $d(J) = k+1$, so $u \in P_k$ by Corollary 1.45. Since it is obvious that $|v|_\alpha$ $(0 < \alpha < 1)$ and $|v|_1^{\wedge}$ vanish if and only if v is constant, i.e. $v \in P_0$, the assertion follows immediately. #

The corresponding assertion for N_α when α is an integer ought to be that $N_\alpha = P_\alpha$. It comes as something of a surprise that this is true only when G is Abelian. In fact, recalling that P_1^{iso} is the set of functions u on G such that $u \circ \exp$ is a first-degreee polynomial on $\mathfrak{g}$, we have:

(5.19) PROPOSITION. *If* α *is a positive integer*,

$$N_\alpha = \{u \in P_{\alpha+1} : X^I u \in P_1^{iso} \quad \text{for} \quad d(I) = \alpha\}.$$

Proof: It suffices to prove the assertion for $\alpha = 1$, namely that $N_1 = P_2 \cap P_1^{iso}$. The proof proceeds in five steps.

Step 1. We first prove the proposition when G is Abelian, i.e. $G = \mathbb{R}^n$. Then $P_1 = P_1^{iso}$, so we must show that $N_1 = P_1$. Writing the group law additively, we have that $u \in N_1$ if and only if

$$u(s+t) + u(s-t) - 2u(s) = 0 \quad \text{for all} \quad s,t \in G. \tag{5.20}$$

It is thus obvious that $P_1 \subset N_1$. On the other hand, suppose $u \in N_1$. Given

$x,y \in G$, we take $s = (x+y)/2$, $t = (x-y)/2$ and then $s = t = (x+y)/2$ in (5.20), obtaining

$$u(x)+u(y) - 2u((x+y)/2) = 0 = u(x+y) + u(0) - 2u((x+y)/2).$$

Therefore, setting $v(x) = u(x) - u(0)$, we have

$$v(x)+v(y) = v(x+y) \quad \text{for all } x,y \in G.$$

Since v is continuous, it follows easily that v is linear, and hence that $u \in P_1$.

Step 2. Returning to the general stratified G, suppose $u \in N_1$. For any $x \in G$ and $Y \in \mathfrak{g}$ the function $f(t) = u(x\cdot\exp(tY))$ satisfies (5.20), so by Step 1 it is of the form $f(t) = u(x)+Ct$. Let us choose $\phi \in C_0^\infty(\mathbb{R})$ with $\int\phi(t)dt = 1$ and $\int t\phi(t)dt = 0$: it follows that

$$(5.21) \qquad \int u(x\cdot\exp(tY))\phi(t)dt = u(x).$$

Let $\Phi : \mathbb{R}^n \to G$ be as in Lemma 1.31, that is,

$$\Phi(t_1,\ldots,t_n) = (\exp(tX_1))(\exp(tX_2))\cdots(\exp(tX_n)),$$

and define $\psi \in C_0^\infty(G)$ by

$$\psi(\Phi(t_1,\ldots,t_n)^{-1}) = \phi(t_1)\phi(t_2)\cdots\phi(t_n)J_\Phi(t_1,\ldots,t_n)^{-1}$$

where J_Φ is the Jacobian determinant of Φ. Then by (5.21),

$$u*\psi(x) = \int u(xy)\psi(y^{-1})dy = \int u(x(\exp(tX_1))\cdots(\exp(tX_n)))\phi(t_1)\cdots\phi(t_n)dt_1\cdots dt_n$$

$$= u(x).$$

Therefore u is C^∞ on G.

Step 3. If $u \in N_1$, by subtracting off a constant we may assume that $u(0) = 0$. Then, as in Step 2, it follows from Step 1 that $u(\exp(tY)) = tu(\exp Y)$ for any $y \in G$. By Step 2, $u \circ \exp$ is differentiable: let u' be its differential at 0. Then for any $Y \in \mathfrak{g}$,

$$u(\exp(tY)) - u(0) - u'(tY) = t[u(\exp Y) - u'(Y)].$$

By definition of u', the quantity on the left is $o(t)$ as $t \to 0$, and this can only happen if $u(\exp Y) = u'(Y)$. Thus $u \circ \exp$ is linear on $\mathfrak{g}$, which means that $u \in P_1^{iso}$.

Step 4. Suppose $u \in N_1$. Then u is smooth by Step 2, and for any $Y \in \mathfrak{g}$, $u(x \cdot \exp(tY)) - u(x)$ is linear in t, hence $Y^2 u = 0$. For any $X, Y \in \mathfrak{g}$, then, we have $X^2 u = Y^2 u = (X+Y)^2 u = 0$, so that

$$(5.22) \qquad (XY+YX)u = 0, \qquad [X,Y]u = 2XYu.$$

Therefore, for any $X, Y, Z \in \mathfrak{g}$,

$$4XYZu = 2X[Y,Z]u = [X,[Y,Z]]u.$$

Applying the Jacobi identity to this equation, we obtain

$$(5.23) \qquad XYZu + ZXYu + YZXu = 0.$$

On the other hand, by (5.22),

$$(5.24) \qquad 2XYZu = \tfrac{1}{2}[X,[Y,Z]]u = \tfrac{1}{2}[[Z,Y],X]u = [Z,Y]Xu = ZYXu - YZXu.$$

Subtracting (5.23) from (5.24) and using (5.22) again,

$$XYZu = ZXYu + ZYXu = Z(XY+YX)u = 0.$$

Since G is stratified, any X^I with $d(I) = 3$ is a linear combination of terms of the form XYZ where X,Y,Z are homogeneous of degree 1. Therefore, $X^I u = 0$ whenever $d(I) = 3$, so $u \in \mathcal{P}_2$ by Corollary 1.45.

<u>Step 5</u>. We have now shown that $\mathcal{N}_1 \subset (\mathcal{P}_2 \cap \mathcal{P}_1^{iso})$. Conversely, suppose that $u \in \mathcal{P}_2 \cap \mathcal{P}_1^{iso}$ and (without loss of generality) that $u(0) = 0$, so that u is a linear combination of the coordinate functions η_j in Chapter 1, Section C, with $d_j \leq 2$. Then by (1.23) and (1.24),

$$u(xy) = u(x)+u(y) + \Sigma P_i(x)Q_i(y)$$

for some $P_i, Q_i \in \mathcal{P}_1$ which vanish at 0. Since $u \circ \exp$ and $Q_i \circ \exp$ are linear,

$$u(y^{-1}) = -u(y), \qquad Q_i(y^{-1}) = -Q_i(y).$$

From these equations it follows immediately that $u \in \mathcal{N}_1$. #

We now return to the Campanato spaces C_N^α. In the following sequence of lemmas we always assume implicitly that $N \leq [\alpha]$. To begin with, we observe that if $u \in C_N^\alpha$, $x_0 \in G$, and $r > 0$, the map

$$P \to \text{ess sup}_{x \in B(r,x_0)} |u(x) - P(x)|$$

is continuous from $\mathcal{P}_N$ to $[0,\infty)$ and tends to ∞ as $\sup_{x \in B(r,x_0)} |P(x)| \to \infty$, so by local compactness of $\mathcal{P}_N$ there exists $P_{r,x_0} \in \mathcal{P}_N$ such that

$$(5.25) \qquad \text{ess sup}_{x \in B(r,x_0)} |u(x) - P_{r,x_0}(x)| = \inf_{P \in \mathcal{P}_N} \text{ess sup}_{x \in B(r,x_0)} |u(x) - P($$

For each $x_0 \in G$ and $r > 0$ we fix once and for all a $P_{r,x_0} \in \mathcal{P}_N$ satisfying (5.25). (P_{r,x_0} also depends on u and N, of course, but this will cause no confusion.)

(5.26) LEMMA. For each $N \in \mathbb{N}$ there exists $C_1 > 0$ such that for all $x_0 \in G$, $r > 0$, $s \geq 1$, and $P \in \mathcal{P}_N$,

$$|X^I P(x_0)| \leq C_1 r^{-d(I)} \sup_{x \in B(r,x_0)} |P(x)| \qquad \text{for } d(I) \leq N,$$

$$\sup_{x \in B(sr,x_0)} |P(x)| \leq C_1 s^N \sup_{x \in B(r,x_0)} |P(x)|.$$

Proof: By making the change of variable $Q(x) = P(x_0(rx))$, it suffices to show that for all $Q \in \mathcal{P}_N$ and $s \geq 1$,

$$|X^I Q(0)| \leq C_1 \sup_{x \in B(1,0)} |Q(x)| \qquad \text{for } d(I) \leq N,$$

$$\sup_{x \in B(s,0)} |Q(x)| \leq C_1 s^N \sup_{x \in B(1,0)} |Q(x)|.$$

But this is clear since $\mathcal{P}_N$ is finite-dimensional and the functions

$$Q \to \sup_{x \in B(1,0)} |Q(x)|, \qquad Q \to \sup_{s \geq 1} \sup_{x \in B(s,0)} s^{-N} |Q(x)|$$

are both norms on $\mathcal{P}_N$. #

(5.27) LEMMA. There exists $C_2 > 0$ such that for all $u \in C_N^\alpha$, $x_0 \in G$, $r > 0$, and $s \geq 1$,

$$\operatorname{ess\,sup}_{x \in B(sr,x_0)} |u(x) - P_{r,x_0}(x)| \leq C_2 \nu_N^\alpha(u) s^\alpha r^\alpha \qquad \text{if } \alpha > N,$$

$$\operatorname{ess\,sup}_{x \in B(sr,x_0)} |u(x) - P_{r,x_0}(x)| \leq C_2 \nu_N^\alpha(u) s^\alpha (1 + \log s) r^\alpha \qquad \text{if } \alpha = N.$$

<u>Proof</u>: Suppose $2^{k-1} \leq s \leq 2^k$. Then by Lemma 5.26,

$$\text{ess sup}_{x\in B(sr,x_0)} |u(x) - P_{r,x_0}(x)|$$

$$\leq \text{ess sup}_{x\in B(2^k r,x_0)} |u(x) - P_{r,x_0}(x)|$$

$$\leq \text{ess sup}_{x\in B(2^k r,x_0)} |u(x) - P_{2^k r,x_0}(x)|$$

$$+ \Sigma_0^{k-1} \text{ ess sup}_{x\in B(2^k r,x_0)} |P_{2^{j+1}r,x_0}(x) - P_{2^j r,x_0}(x)|$$

$$\leq \nu_N^\alpha(u)(2^k r)^\alpha$$

$$+ C_1\Sigma_0^{k-1} 2^{(k-j)N} \text{ ess sup}_{x\in B(2^j r,x_0)} |P_{2^{j+1}r,x_0}(x) - P_{2^j r,x_0}(x)|$$

$$\leq \nu_N^\alpha(u)(2^k r)^\alpha$$

$$+ C_1\Sigma^{k-1} 2^{(k-j)N} \text{ ess sup}_{x\in B(2^j r,x_0)} [|P_{2^{j+1}r,x_0}(x)-u(x)|+|u(x)-P_{2^j r,x_0}($$

$$\leq \nu_N^\alpha(u)[(2^k r)^\alpha + C_1\Sigma_0^{k-1} 2^{(k-j)N}[(2^{j+1}r)^\alpha + (2^j r)^\alpha]]$$

$$\leq \nu_N^\alpha(u) r^\alpha [2^{k\alpha} + C_1 2^{\alpha+kN}\Sigma_0^{k-1} 2^{j(\alpha-N)}].$$

Thus if $\alpha > N$, by summing the series we obtain

$$\operatorname{ess\,sup}_{x\in B(sr,x_0)} |u(x) - P_{r,x_0}(x)|$$

$$\le \nu_N^\alpha(u) r^\alpha [2^{k\alpha} + C_1 2^{\alpha+kN} 2^{k(\alpha-N)} (2^{\alpha-N} - 1)^{-1}]$$

$$\le [2^\alpha + C_1 2^{2\alpha}(2^{\alpha-N} - 1)^{-1}] \nu_N^\alpha(u) r^\alpha 2^{(k-1)\alpha}$$

$$\le C_2 \nu_N^\alpha(u) r^\alpha s^\alpha,$$

whereas if $\alpha = N$,

$$\operatorname{ess\,sup}_{x\in B(sr,x_0)} |u(x) - P_{r,x_0}(x)| \le \nu_N^\alpha(u) r^\alpha [2^{k\alpha} + 2^{\alpha+k\alpha} C_1 k]$$

$$\le C_2 \nu_N^\alpha(u) r^\alpha s^\alpha (1 + \log s). \quad \#$$

(5.28) LEMMA. *If* $u \in C_N^\alpha$, *then*

$$|u(x)| = O(1 + |x|^\alpha) \qquad \text{a.e. } \textit{if } \alpha > N,$$

$$|u(x)| = O((1 + |x|^\alpha)\log(2 + |x|)) \quad \text{a.e. } \textit{if } \alpha = N.$$

Proof: In Lemma 5.27, take $r = 1$, $x_0 = 0$, and use the fact that $|P(x)| = O(1 + |x|^N)$ for all $P \in \mathcal{P}_N$.

(5.29) LEMMA. *Let* ϕ *be a measurable function on* G *such that*

(a) $|\phi(x)| = O(1 + |x|)^{-\beta}$ *where* $\beta > Q + \alpha$;

(b) $\int P(x)\phi(x)dx = 0$ *for all* $P \in \mathcal{P}_N$.

<u>Then there exists</u> $C_3 > 0$ <u>such that for all</u> $u \in C_N^\alpha$ <u>and</u> $t > 0$,

$$\|u*\phi_t\|_\infty \le C_3 \nu_N^\alpha(u) t^\alpha.$$

<u>Proof</u>: Condition (a) and Lemma 5.28 ensure that the integrals $\int P(x)\phi(x)dx$ and $u*\phi_t(x)$ all converge absolutely. By condition (b),

$$u*\phi_t(x) = \int [u(xy^{-1}) - P_{t,x}(xy^{-1})]\phi_t(y)dy$$

We break up this integral as

$$\int_G = \int_{|y|\le t} + \Sigma_1^\infty \int_{2^{j-1}t\le|y|\le 2^j t}$$

and use Lemma 5.27, obtaining for $\alpha > N$:

$$\begin{aligned}
|u*\phi_t(x)| &\le \nu_N^\alpha(u)t^\alpha\Big[\int_{|y|\le t} |\phi_t(y)|dy + \Sigma_1^\infty C_2 2^{j\alpha}\int_{2^{j-1}t\le|y|\le 2^j t} |\phi_t(y)|dy\Big] \\
&= \nu_N^\alpha(u)t^\alpha\Big[\int_{|y|\le 1} |\phi(y)|dy + \Sigma_1^\infty C_2 2^{j\alpha}\int_{2^{j-1}\le|y|\le 2^j} |\phi(y)|dy\Big] \\
&\le \nu_N^\alpha(u)t^\alpha C\Big[\int_{|y|\le 1} dy + \Sigma_1^\infty C_2 2^{j\alpha}\int_{|y|>2^{j-1}} |y|^{-\beta}dy\Big] \\
&\le \nu_N^\alpha(u)t^\alpha C[1 + C'\Sigma_1^\infty 2^{j(\alpha+Q-\beta)}].
\end{aligned}$$

Since $\beta > Q+\alpha$ the assertion for $\alpha > N$ is proved. If $\alpha = N$ the proof is the same except that the terms $2^{j\alpha}$ must be replaced by $Cj2^{j\alpha}$. #

(5.30) LEMMA. <u>Let</u> $h(x,t)$, H_t <u>be the heat kernel and heat semigroup defined in Chapter</u> 1, <u>Section</u> G. <u>If</u> $k \in \mathbb{N}$ <u>and</u> $d(J) > N-2k$ <u>there exists</u> $C_4 > 0$ <u>such that for all</u> $u \in C_N^\alpha$ <u>and</u> $t > 0$,

$$\|X^J \partial_t^k H_t u\|_\infty \le C_4 \nu_N^\alpha(u) t^{(\alpha-2k-d(J))/2}.$$

Proof: Let

$$\phi(x) = X^J \partial_t^k h(x,t)|_{t=1} = X^J(-L)^k h(x,1).$$

where L is the sub-Laplacian. Then ϕ satisfies the hypotheses of Lemma 5.29: (a) is obvious since $h(\cdot,1) \in S$, and (b) is true since for any $P \in P_N$,

$$\int P(x)\phi(x)dx = \pm\int (L^k X^J P)(x)h(x,1)dx = \int 0dx = 0$$

Moreover, by (1.73),

$$\phi_t(x) = t^{-Q} X^J(-L)^k h(x/t,1) = t^{d(J)+2k} X^J(-L)^k h(x,t^2),$$

so that

$$X^J \partial_t^k h(x,t) = t^{-(d(J)+2k)/2}\ \phi_{\sqrt{t}}(x).$$

The assertion thus follows immediately from Lemma 5.29. #

(5.31) LEMMA. *If* $u \in C_N^\alpha$ *then, after modification on a set of measure zero,* $u \in C^{[\alpha]}$ *when* $\alpha > N$ *and* $u \in C^{N-1}$ *when* $\alpha = N$.

Proof: First, let k be the integer such that $(\alpha/2) < k \leq (\alpha/2)+1$. By Lemma 5.30, $\|\partial_t^k H_t u\|_\infty = O(t^{(\alpha/2)-k})$. Integrating $k-1$ times in t, we find that as $t \to 0$, $\|\partial_t H_t u\|_\infty$ is $O(t^{(\alpha/2)-1})$ if $\alpha < 2$, $O(|\log t|)$ if $\alpha = 2$, and $O(1)$ if $\alpha > 2$. In any case, $\|\partial_t H_t u\|_\infty$ is integrable on $[0,1]$. Since $H_t u \to u$ in S', we have

$$u = H_1 u - \lim_{\varepsilon \to 0} \int_\varepsilon^1 \partial_t H_t u dt.$$

But the limit on the right exists in the uniform norm, and $H_t u \in C^\infty$. Therefore we can correct u on a set of measure zero to make it continuous. Once this is done, we have

$$(5.32)\qquad (H_1 - I)^k u = \int_0^1 \cdots \int_0^1 (\partial_t H_{t_1}) \cdots (\partial_t H_{t_k}) u\, dt_1 \cdots dt_k$$

(where by $\partial_t H_a$ we mean $\partial_t H_t|_{t=a}$), the integral converging uniformly. But also

$$(H_1 - I)^k u = (-1)^k u + \Sigma_0^{k-1} (-1)^j \binom{k}{j} H_{k-j} u.$$

Since $H_{k-j}u$ is C^∞, to prove that $X^J u$ is continuous it suffices to prove that $X^J(H_1 - I)^k u$ is continuous, and this we shall do by differentiating under the integral in (5.32). Observe that since $H_{t_1} \cdots H_{t_k} = H_{t_1 + \cdots + t_k}$,

$$(\partial_t H_{t_1}) \cdots (\partial_t H_{t_k}) = \partial_{t_1} \cdots \partial_{t_k} (H_{t_1 + \cdots + t_k}) = \partial_t^k H_{t_1 + \cdots + t_k}.$$

Therefore, by Lemma 5.30 we have

$$\int_0^1 \cdots \int_0^1 \| X^J (\partial_t H_{t_1}) \cdots (\partial_t H_{t_k}) u \|_\infty \, dt_1 \cdots dt_k$$

$$= \int_0^1 \cdots \int_0^1 \| X^J \partial_t^k H_{t_1 + \cdots + t_k} u \|_\infty \, dt_1 \cdots dt_k$$

$$\leq C_4 \nu_N^\alpha(u) \int_0^1 \cdots \int_0^1 (t_1 + \cdots t_k)^{(\alpha - 2k - d(J))/2} \, dt_1 \cdots dt_k,$$

and the last integral converges provided that $(\alpha-2k-d(J))/2 > -k$, that is, provided that $d(J) < \alpha$. Hence for such J's the integral

$$\int_0^1 \cdots \int_0^1 X^J (\partial_t H_{t_1}) \cdots (\partial_t H_{t_k}) u\, dt_1 \cdots dt_k$$

converges absolutely and uniformly, so we can interchange differentiation and integration to conclude that $X^J u$ is continuous for $d(J) < \alpha$. #

We return to the study of the polynomials P_{r,x_0}. If $u \in C_N^\alpha$ we may assume that u is continuous by Lemma 5.31, so henceforth we shall replace "ess sup" by "sup" in (5.25) and similar expressions. We define

$$a_I(r,x_0) = X^I P_{r,x_0}(x_0),$$

and observe that $a_I(r,x_0) = 0$ if $d(I) > N$.

(5.33) LEMMA. <u>There exists</u> $C_5 > 0$ <u>such that for all</u> $u \in C_N^\alpha$, $r > 0$, <u>and</u> $x_0, y_0 \in G$ <u>with</u> $|x_0^{-1}y_0| \leq r$,

$$|a_I(2\gamma r,x_0) - a_I(2\gamma r,y_0)| \leq C_5 \nu_N^\alpha(u) r^{\alpha-N} \quad \text{when } d(I) = N.$$

<u>Proof</u>: If $d(I) = N$ then $X^I P$ is constant for all $P \in \mathcal{P}_N$. Hence by Lemma 5.26, since $B(r,x_0) \subset B(2\gamma r,y_0)$,

$$|a_I(2\gamma r,x_0) - a_I(2\gamma r,y_0)| = |X^I(P_{2\gamma r,x_0} - P_{2\gamma r,y_0})|$$

$$\leq C_1 r^{-N} \sup_{x\in B(r,x_0)} |P_{2\gamma r,x_0}(x) - P_{2\gamma r,y_0}(x)|$$

$$\leq C_1 r^{-N}[\sup_{x\in B(2\gamma r,x_0)} |P_{2\gamma r,x_0}(x) - u(x)| + \sup_{x\in B(2\gamma r,y_0)} |u(x) - P_{2\gamma r,y_0}(x)|]$$

$$\leq 2^{\alpha+1}\gamma^\alpha C_1 \nu_N^\alpha(u) r^{\alpha-N}. \quad \#$$

(5.34) LEMMA. <u>There exists</u> $C_6 > 0$ <u>such that for all</u> $u \in C_N^\alpha$, $r > 0$, $0 < \varepsilon < 1$, <u>and</u> $x_0 \in G$,

$$|a_I(r,x_0) - a_I(\varepsilon r,x_0)| \leq C_6 \nu_N^\alpha(u) r^{\alpha - d(I)} \qquad \text{if } d(I) < \alpha,$$

$$|a_I(r,x_0) - a_I(\varepsilon r,x_0)| \leq C_6 \nu_N^\alpha(u)(1 + |\log \varepsilon|) \quad \text{if } d(I) = \alpha = N.$$

<u>Proof</u>: We may suppose that $d(I) \leq N$, since otherwise $a_I = 0$. First, suppose that $2^{-1} \leq \varepsilon < 1$. By Lemma 5.26,

$$|a_I(r,x_0) - a_I(\varepsilon r,x_0)| = |X^I(P_{r,x_0} - P_{\varepsilon r,x_0})(x)|$$

$$\leq C_1(\varepsilon r)^{-d(I)} \sup_{x \in B(\varepsilon r,x_0)} |P_{r,x_0} - P_{\varepsilon r,x_0}(x)|$$

$$\leq C_1(\varepsilon r)^{-d(I)} \sup_{x \in B(\varepsilon r,x_0)} [|P_{r,x_0}(x) - u(x)| + |u(x) - P_{\varepsilon r,x_0}(x)|]$$

$$\leq C_1(\varepsilon r)^{-d(I)} \nu_N^\alpha(u)[r^\alpha + (\varepsilon r)^\alpha]$$

$$\leq C_1 2^{N+1} \nu_N^\alpha(u) r^{\alpha - d(I)}.$$

Now suppose that $2^{-(k+1)} \leq \varepsilon < 2^{-k}$. Then by the result for $2^{-1} \leq \varepsilon < 1$,

$$|a_I(r,x_0) - a_I(\varepsilon r,x_0)|$$

$$\leq \Sigma_0^{k-1} |a_I(2^{-j}r,x_0) - a_I(2^{-(j+1)}r,x_0)| + |a_I(2^{-k}r,x_0) - a_I(\varepsilon r,x_0)|$$

$$\leq C_1 2^{N+1} \nu_N^\alpha(u) r^{\alpha - d(I)} \Sigma_0^k 2^{-j(\alpha - d(I))}.$$

If $d(I) < \alpha$ the last sum is less than $(1 - 2^{d(I)-\alpha})^{-1}$, whereas if $d(I) = \alpha$ it equals $k+1$, which is comparable to $1 + |\log \varepsilon|$. #

(5.35) LEMMA. <u>If</u> $u \in C_N^\alpha$ <u>then whenever</u> $d(I) < \alpha$,

$$\lim_{r\to 0} a_I(r,x_0) = X^I u(x_0)$$

<u>for all</u> $x_0 \in G$. <u>Moreover</u>,

$$(5.36) \qquad |a_I(r,x_0) - X^I u(x_0)| \le C_6 \nu_N^\alpha(u) r^{\alpha - d(I)}.$$

<u>Proof</u>: Let $v_I(x_0) = \lim_{r\to 0} a_I(r,x_0)$. The existence of $X^I u(x_0)$ and $v_I(x_0)$ is guaranteed by Lemmas 5.31 and 5.34, and the estimate (5.36) will follow from Lemma 5.34 once we have shown that $v_I(x_0) = X^I u(x_0)$.

Suppose to the contrary that there exist $x_0 \in G$ and J with $d(J) < \alpha$ such that $v_J(x_0) \ne X^J u(x_0)$. Let k be the integer such that $k < \alpha \le k+1$, and let P be the left Taylor polynomial of u at x_0 of homogeneous degree k. Also, let Q_{r,x_0} be the left Taylor polynomial of P_{r,x_0} at x_0 of homogeneous degree k: thus $Q_{r,x_0}(x) = P_{r,x_0}(x_0 x)$ unless $\alpha = N = k+1$, in which case $Q_{r,x_0}(x)$ is obtained from $P_{r,x_0}(x_0 x)$ by omitting the terms which are homogeneous of degree N. Then $X^I Q_{r,x_0}(0) = a_I(r,x_0)$ for $d(I) \le k$, and it follows that Q_{r,x_0} converges in $\mathcal{P}_k$ as $r \to 0$ to a polynomial $Q \in \mathcal{P}_k$ such that $X^I Q(0) = v_I(x_0)$ for $d(I) \le k$. Since $X^J Q(0) \ne X^J P(0)$ there exists $r_1 \le 1$ and $C_7 > 0$ such that

$$\sup_{x\in B(r,0)} |P(x) - Q(x)| \ge 2C_7 r^{d(J)} \ge 2C_7 r^k \qquad \text{for } r \le r_1.$$

Also, from the stratified Taylor inequality (1.42) it follows that there exists $r_2 \le r_1$ such that

$$\sup_{x\in B(r,0)} |u(x_0 x) - P(x)| \le C_7 r^k \qquad \text{for } r \le r_2,$$

and hence

$$\sup_{x\in B(r,0)} |u(x_0x) - Q(x)| \geq C_7 r^k \quad \text{for} \quad r \leq r_2. \tag{5.37}$$

On the other hand, if $\alpha > N$,

$$\sup_{x\in B(r,0)} |u(x_0x) - Q_{r,x_0}(x)| = \sup_{x\in B(r,0)} |u(x_0x) - P_{r,x_0}(x_0x)| \leq \nu_N^\alpha(u)r^\alpha,$$

while if $\alpha = N$, by Lemma 5.34 we have for $r \leq 1/2$:

$$\sup_{x\in B(r,0)} |u(x_0x) - Q_{r,x_0}(x)|$$

$$\leq \sup_{x\in B(r,0)} |u(x_0x) - P_{r,x_0}(x_0x)| + \sup_{x\in B(r,0)} |P_{r,x_0}(x_0x) - Q_{r,x_0}(x)|$$

$$\leq \nu_N^\alpha(u)r^\alpha + C_8\Sigma_{d(I)=k+1} |a_I(r,x_0)|r^{k+1}$$

$$\leq \nu_N^\alpha(u)[r^\alpha + C_8C_6|\log r|r^{k+1}]$$

$$\leq C_9\nu_N^\alpha(u)r^\alpha|\log r|.$$

Moreover, by Lemma 5.34,

$$\sup_{x\in B(r,0)} |Q_{r,x_0}(x) - Q(x)| \leq C_{10}\Sigma_{d(I)\leq k} |a_I(r,x_0) - v_I(x_0)|r^{d(I)}$$

$$\leq C_6C_{10}\Sigma_{d(I)\leq k}\, r^{\alpha-d(I)}r^{d(I)}\nu_N^\alpha(u)$$

$$= C_{11}\nu_N^\alpha(u)r^\alpha.$$

Therefore, no matter whether $\alpha < N$ or $\alpha = N$,

$$(5.38) \quad \sup_{x\in B(r,0)} |u(x_0x) - Q(x)| \leq C_{12}\nu_N^\alpha(u)r^\alpha|\log r| \quad \text{for } r \leq 1/2.$$

Since $k < \alpha$, (5.37) and (5.38) are inconsistent for sufficiently small r, so our hypothesis $v_J(x_0) \neq X^J u(x_0)$ has led to a contradiction. #

We now come to the main theorem toward which we have been working, which, in conjunction with Corollary 5.5, characterizes C_N^α as a Lipschitz class in all cases except when α is an integer and $N \geq \alpha$.

(5.39) THEOREM. *Suppose* $N \in \mathbb{N}$ *and* $\alpha > N$.

(a) *If* $N < \alpha < N+1$ *then* $C_N^\alpha = \Gamma_\alpha^{hom}$.

(b) *If* $\alpha = N+1$ *then* $C_N^\alpha = \hat{\Gamma}_\alpha^{hom}$.

(c) *If* $\alpha > N+1$ *then* $C_N^\alpha = P_N$.

In cases (a) *and* (b) *the seminorms* ν_N^α *and* $|\ |_\alpha$ (*or* $|\ |_\alpha^\wedge$) *are equivalent*.

Proof: Suppose $N < \alpha \leq N+1$. By Lemmas 5.33 and 5.35, if $u \in C_N^\alpha$, $d(I) = N$, and $x,y \in G$ we have

$$\begin{aligned} |X^I u(xy) - X^I u(x)| &\leq |X^I u(xy) - a_I(2\gamma|y|,xy)| + |a_I(2\gamma|y|,xy) - a_I(2\gamma|y|,x)| \\ &\quad + |a_I(2\gamma|y|,x) - X^I u(x)| \\ &\leq \nu_N^\alpha(u)[C_6(2\gamma|y|)^{\alpha-N} + C_5|y|^{\alpha-N} + C_6(2\gamma|y|)^{\alpha-N}] \\ &= C_{13}\nu_N^\alpha(u)|y|^{\alpha-N}, \end{aligned}$$

so that $u \in \Gamma_\alpha^{hom}$ if $\alpha < N+1$ and $u \in \hat{\Gamma}_\alpha^{hom}$ if $\alpha = N+1$; moreover, $|u|_\alpha \leq C_{13}\nu_N^\alpha(u)$ or $|u|_\alpha^\wedge \leq C_{13}\nu_N^\alpha(u)$ respectively. Conversely, if $u \in \Gamma_\alpha^{hom}$ $(\alpha \notin \mathbb{N})$ or $u \in \hat{\Gamma}_\alpha^{hom}$ $(\alpha \in \mathbb{N})$ then for $N = [\alpha]$ or $N = \alpha-1$ respectively, and $x_0 \in G$, let $P_{x_0} \in \mathcal{P}_N$ be the left Taylor polynomial of u at x_0 of homogeneous degree N, and let $Q_{x_0}(x) = P_{x_0}(x_0^{-1}x)$. Then by the stratified Taylor inequality (1.42),

$$\nu_N^\alpha(u) \leq \sup_{x_0 \in G, r>0} \sup_{x \in B(r,x_0)} r^{-\alpha}|u(x) - Q_{x_0}(x)|$$

$$\leq C_{14}|u|_\alpha \qquad \text{or} \qquad C_{14}|u|_\alpha^\wedge .$$

This proves (a) and (b). Finally, if $\alpha > N+1$ then Lemma 5.35 implies that if $u \in C_N^\alpha$ then $X^I u$ vanishes indentically for $d(I) = N+1$, and hence $u \in \mathcal{P}_N$ by Corollary 1.45. #

It now remains to characterize C_N^α when α is a positive integer and $N \geq \alpha$. As before, it suffices to consider $N = \alpha$.

(5.40) THEOREM. <u>If N is a positive integer, then $C_N^N \subset \Gamma_N^{hom}$ and ν_N^N dominates $|\ |_N$.</u>

<u>Proof</u>: If $u \in C_N^N$, Lemma 5.35 implies that for $x,y \in G$, $r = 4\gamma^2|y|$, and $d(I) = N-1$,

$$|X^I u(xy) + X^I u(xy^{-1}) - 2X^I u(x)|$$

$$\leq |a_I(r,xy) + a_I(r,xy^{-1}) - 2a_I(r,x)| + |X^I u(xy) - a_I(r,xy)|$$

$$+ |X^I u(xy^{-1}) - a_I(r,xy^{-1})| + 2|X^I u(x) - a_I u(x)|$$

$$\leq |a_I(r,xy) + a_I(r,xy^{-1}) - 2a_I(r,x)| + 16\gamma^2 C_6 \nu_N^N(u)|y|$$

$$\leq 16\gamma^2 C_6 \nu_N^N(u)|y| + |X^I(P_{r,xy} + P_{r,xy^{-1}} - 2P_{r,x})(x)|$$

$$+ |X^I P_{r,xy}(xy) - X^I P_{r,xy}(x) + X^I P_{r,xy^{-1}}(xy^{-1}) - X^I P_{r,xy^{-1}}(x)|$$

$$= 16\gamma^2 C_6 \nu_N^N(u)|y| + T_1 + T_2.$$

We estimate T_1 as in Lemma 5.33:

$$T_1 \leq C_1 |y|^{1-N} \sup_{z\in B(|y|,x)} |(P_{r,xy} + P_{r,xy^{-1}} - 2P_{r,x})(z)|$$

$$\leq C_1 |y|^{1-N} \sup_{z\in B(|y|,x)} [|u(z)-P_{r,xy}(z)| + |u(z)-P_{r,xy^{-1}}(z)| + 2|u(z)-P_{r,x}(z)|]$$

$$\leq C_1 |y|^{1-N} 4\nu_N^N(u) r^N$$

$$= 4^{N+1}\gamma^{2N} C_1 \nu_N^N(u)|y|.$$

To handle T_2, we make the following observation: for any $Q \in \mathcal{P}_1$, $Q(xy) - Q(x)$ depends only on y and not on x. Indeed, this follows from (1.23) since Q is a linear combination of 1 and the coordinate functions η_j with $d_j = 1$. Taking $Q = X^I P_{r,xy^{-1}}$, then, we see that

$$X^I P_{r,xy^{-1}}(xy^{-1}) - X^I P_{r,xy^{-1}}(x) = X^I P_{r,xy^{-1}}(x) - X^I P_{r,xy^{-1}}(xy).$$

Thus, by the stratified mean value theorem (1.41),

$$T_2 \leq |X^I(P_{r,xy} - P_{r,xy^{-1}})(xy) - X^I(P_{r,xy} - P_{r,xy^{-1}})(x)|$$

$$\leq C|y| \sup_{|z|\leq b|y|,\, 1\leq j\leq \nu} |X_j X^I(P_{r,xy} - P_{r,xy^{-1}})(z)|.$$

But $X_j X^I$ is a linear combination of X^K's with $d(K) = N$, and $X^K P$ is a constant function for $d(K) = N$ and $P \in \mathcal{P}_N$, so by Lemma 5.33,

$$T_2 \leq C'|y| \Sigma_{d(K)=N} |a_K(r,xy) - a_K(r,xy^{-1})|$$

$$\leq C''|y| \nu_N^N(u).$$

Therefore, finally,

$$|X^I u(xy) + X^I u(xy^{-1}) - 2X^I u(x)| \leq [16\gamma^2 C_6 + 4^{N+1}\gamma^{2N} C_1 + C''] \nu_N^N(u)|y|,$$

and we are done. #

The reverse inclusion $\Gamma_N^{hom} \subset C_N^N$ is more problematical. Indeed, this relation is definitely <u>false</u> when G is non-Abelian: by Lemma 5.28, any $u \in C_N^N$ satisfies $|u(x)| = O(|x|^N \log|x|)$ as $x \to \infty$, whereas by Proposition 5.19 there exist elements of Γ_N^{hom} which grow like $|x|^{N+1}$ as $x \to \infty$. However, we shall now show that $\Gamma_N^{hom} \subset C_N^N$ when G is Abelian. In the following lemma and theorem we take $G = \mathbb{R}^n$ and write the group law additively, but continue to use the notation X^I for left-invariant (i.e., constant-coefficient) differential monomials.

(5.41) LEMMA. <u>If</u> $u \in \Gamma_1^{hom}(\mathbb{R}^n)$ <u>then</u> $|u(x)| = O(|x|\log|x|)$ <u>as</u> $x \to \infty$.

<u>Proof</u>: Let $v(x) = u(x) - u(0)$. Then for all $x \in \mathbb{R}^n$,

$$|v(2x) - 2v(x)| = |u(2x) + u(0) - 2u(x)| \leq |u|_1 |x|.$$

Setting $x = 2^j y$, we have

$$|2^{-j}v(2^j y) - 2^{-j-1}v(2^{j+1}y)| = 2^{-j-1}|v(2^{j+1}y) - 2v(2^j y)|$$

$$\leq 2^{-j-1}|u|_1|2^j y| = |u|_1|y|/2.$$

Hence for any positive integer k,

$$|v(y) - 2^{-k}v(2^k y)| \leq \Sigma_0^{k-1} |2^{-j}v(2^j y) - 2^{-j-1}v(2^{j+1}y)| \leq k|u|_1|y|/2,$$

so that

$$|v(2^k y)| \leq 2^{k-1}k|u|_1|y| + 2^k|v(y)|.$$

Setting $C = \sup_{|y|\leq 1} |v(y)|$ and $z = 2^k y$, we obtain

$$\sup_{|z|\leq 2^k} |v(z)| \leq 2^{k-1}k|u|_1 + 2^k C.$$

Therefore, if $r \geq 1$, we take k to be the integer such that $2^{k-1} \leq r < 2^k$ and conclude that

$$\sup_{|z|\leq r} |u(z)| \leq |u(0)| + u_1 r \log_2 r + 2Cr = O(r \log r). \quad \#$$

(5.42) THEOREM. _If_ N _is a positive integer, then_ $\Gamma_N^{hom}(\mathbb{R}^n) \subset C_N^N(\mathbb{R}^n)$, _and_ $|\ |_N$ _dominates_ ν_N^N.

Proof: First we prove the theorem for $N = 1$, and for this purpose we use the heat semigroup as in Lemmas 5.30 and 5.31. If $u \in \Gamma_N^{hom}$, then since $h(x,t) = h(-x,t)$ and $\int \partial_t h(x,t)dx = \partial_t \int h(x,t)dx = \partial_t 1 = 0$, by (1.73) we have

$$(5.43)\qquad |\partial_t H_t u(x)| = \tfrac{1}{2}\Big|\int [u(x+y) + u(x-y) - 2u(x)]\partial_t h(y,t)dy\Big|$$

$$\le C_1|u|_1 \int |y|(|y| + \sqrt{t})^{-n-2}dy$$

$$\le C_1|u|_1[\int_{|y|\le\sqrt{t}} |y|t^{-(n+2)/2}dy + \int_{|y|>\sqrt{t}} |y|^{-n-1}dy]$$

$$\le C_2|u|_1 t^{-1/2}.$$

Therefore,

$$(5.44)\qquad \|H_t u - u\|_\infty = \Big\|\int_0^t \partial_s H_s u ds\Big\|_\infty \le C_2|u|_1 \int_0^t s^{-1/2}ds \le C_3|u|_1 t^{1/2}.$$

Next if $d(I) = 2 \;\; (= |I|)$,

$$\partial_t X^I H_t u = \partial_t X^I(H_{t/2}u*h(\cdot,t/2)) = \partial_t H_{t/2}u*X^I h(\cdot,t/2),$$

so that by (5.43) and Proposition 1.75,

$$(5.45)\qquad \|\partial_t X^I H_t u\|_\infty \le \|\partial_t H_{t/2}u\|_\infty \, \|X^I h(\cdot,t/2)\|_1$$

$$\le C_4|u|_1 t^{-1/2}t^{-1} = C_4|u|_1 t^{-3/2}.$$

Also, by Lemma 5.41 and (1.73),

$$|X^I H_t u(x)| = \Big|\int u(x-y)X^I h(y,t)dy\Big|$$

$$\le C_5 \int (|x-y| + 1)\log(|x-y| + 2)\cdot(|y| + \sqrt{t})^{-n-2}dy$$

which is finite and tends to zero as $t \to \infty$, so that

$$X^I H_t u(x) = -\int_t^\infty \partial_s X^I H_s u(x)ds.$$

Hence by (5.45),

$$(5.46) \qquad \|X^I H_t u(x)\|_\infty \le \int_t^\infty C_4 |u|_1 s^{-3/2} ds = C_6 |u|_1 t^{-1/2} \qquad \text{for } d(I) = 2.$$

Now, given $x_0 \in \mathbb{R}^n$ and $r > 0$, let P_{r,x_0} be the Taylor polynomial of $H_{r^2}u$ at x_0 of degree 1. Then by (5.44), (5.46), and Taylor's theorem,

$$\sup_{x \in B(r,x_0)} |u(x) - P_{r,x_0}(x - x_0)|$$

$$\le \sup_{x \in B(r,x_0)} [|u(x) - H_{r^2}u(x)| + |H_{r^2}u(x) - P_{r,x_0}(x - x_0)|]$$

$$\le C_3 |u|_1 (r^2)^{1/2} + C_6 |u|_1 (r^2)^{-1/2} r^2$$

$$\le C_7 |u|_1 r,$$

so that $u \in C_1^1$ and $\nu_1^1(u) \le C_7 |u|_1$.

Now suppose $N > 1$. If $u \in \Gamma_N^{hom}$ and $d(I) = N-1$, we apply the preceding arguments to $X^I u \in \Gamma_1^{hom}$, obtaining

$$\sup_{x \in B(r,x_0)} |X^I u(x) - P_{r,x_0}^I (x - x_0)| \le C_7 |u|_N r,$$

where P_{r,x_0}^I is the Taylor polynomial of $H_{r^2} X^I u$ at x_0 of degree 1. But $H_{r^2} X^I u = X^I H_{r^2} u$, so $P_{r,x_0}^I = X^I P_{r,x_0}$ where P_{r,x_0} is the Taylor polynomial of $H_{r^2}u$ at x_0 of degree N. Therefore,

$$(5.47)\quad \sup_{d(I)=N-1} \sup_{x\in B(r,x_0)} |X^I u(x) - X^I P_{r,x_0}(x-x_0)| \le C_7 |u|_N r.$$

Let Q_{r,x_0} be the Taylor polynomial of $u - P_{r,x_0}((\cdot)-x_0)$ at x_0 of degree $N-2$: then by (5.47) and Taylor's theorem,

$$\sup_{x\in B(r,x_0)} |u(x) - P_{r,x_0}(x-x_0) - Q_{r,x_0}(x-x_0)| \le C_8 |u|_N r \cdot r^{N-1} = C_8 |u|_N r^N.$$

Since $P_{r,x_0} + Q_{r,x_0} \in \mathcal{P}_N$, we conclude that $u \in C_N^N$ and $\nu_N^N(u) \le C_8 |u|_N$. #

Now, what can we say when G is non-Abelian? The obvious conjecture is that Γ_N^{hom} and C_N^N are the same modulo polynomials; more precisely, with reference to Proposition 5.19,

(5.48) CONJECTURE. $\Gamma_N^{hom} = C_N^N + \mathcal{N}_N$.

We have not found a proof of this, but we offer the following two propositions as supporting evidence.

(5.49) PROPOSITION. <u>Suppose</u> $u \in \Gamma_1^{hom}$. <u>Then the following are equivalent</u>: (a) $|u(x)| = O(|x|\log|x|)$ as $x \to \infty$, (b) $|u(x)| = o(|x|^2)$ as $x \to \infty$, (c) $u \in C_1^1$.

<u>Proof</u>: (c) $\Longrightarrow$ (a) by Lemma 5.28, (a) $\Longrightarrow$ (b) trivially, and the proof of Theorem 5.42 for the case $N = 1$, with the classical Taylor theorem replaced by Corollary 1.44, goes through in the non-Abelian case to show that (b) $\Longrightarrow$ (c). (The place where the hypothesis (b) is used in this proof is in showing that

$$X^I H_t u(x) = -\int_t^\infty \partial_s X^I H_s u(x)\,ds \qquad (d(I) = 2).$$

That (b) is necessary for this equation to hold can be seen by taking u to be an element of $\mathcal{N}_1$ which is homogeneous of degree 2. Then $X^2u = 0$ for all $X \in \mathfrak{g}$, so in particular $\mathcal{L}u = 0$. Thus u is a steady-state solution of $(\partial_t + \mathcal{L})u = 0$, that is, $H_t u = u$ for all t. Hence $X^I H_t u = X^I u$ is constant (in general, non-zero) for $d(I) = 2$, while $\partial_t X^I H_t u = 0$.) #

The proof of Theorem 5.42 does not yield a corresponding result for $N > 1$ in the non-Abelian case, since in general $X^I H_{r^2} u \neq H_{r^2} X^I u$. However, we have the following weaker result.

(5.50) PROPOSITION. <u>If</u> N <u>is a positive integer and</u> $u \in \Gamma_N$ (<u>i.e.</u>, $u \in \Gamma_N^{\mathrm{hom}}$ <u>and</u> $X^I u$ <u>is bounded for</u> $d(I) \leq N-1$), <u>then</u> $u \in C_N^N$.

<u>Proof</u>: By a theorem of Folland [3], if $u \in \Gamma_N$ then for every $r > 0$ there exist $u_r \in \Gamma_{N-(1/2)}$ and $u^r \in \Gamma_{N+(1/2)}$ such that $u = u_r + u^r$, $|u_r|_{N-(1/2)} \leq C_0 r^{1/2}$, and $|u^r|_{N+(1/2)} \leq C_0 r^{-1/2}$, where C_0 is comparable to $|u|_N$. By Theorem 5.39, for each $x_0 \in G$ there exist $P_{r,x_0} \in \mathcal{P}_{N-1}$ and $P^{r,x_0} \in \mathcal{P}_N$ such that

$$\sup_{x \in B(r,x_0)} |u_r(x) - P_{r,x_0}(x)| \leq C|u_r|_{N-(1/2)} r^{N-(1/2)} \leq CC_0 r^N,$$

$$\sup_{x \in B(r,x_0)} |u^r(x) - P^{r,x_0}(x)| \leq C|u^r|_{N+(1/2)} r^{N+(1/2)} \leq CC_0 r^N.$$

But then $P_{r,x_0} + P^{r,x_0} \in \mathcal{P}_N$ and we have

$$\sup_{x \in B(r,x_0)} |u(x) - P_{r,x_0}(x) - P^{r,x_0}(x)| \leq 2CC_0 r^N,$$

so that $u \in C_N^N$. #

Notes and References

The space $BMO(\mathbb{R}^n)$ was introduced by John and Nirenberg [1], who proved Theorem 5.15 in this setting. The first version of Theorem 5.10, the duality in terms of the sharp function, but in the context of dyadic martingales, is an unpublished result of R. Fefferman. Our proof of Theorem 5.10 is adapted from the $\mathbb{R}^n$ version given in Stein [6]. For $G = \mathbb{R}^n$ the theorem that $(H^1)^* = BMO/\mathcal{P}_0$ is announced in C. Fefferman [1]; two proofs are given in C. Fefferman and Stein [1]. Carleson [1] has given yet another proof, and Strömberg [1] has generalized the theorem to spaces defined by Orlicz norms.

The identification of $(H^p)^*$ $(0 < p < 1)$ with a Lipschitz space was first obtained for H^p spaces of holomorphic functions on the disc by Duren, Romberg, and Shields [1]. For $G = \mathbb{R}^n$ this theorem is due to Walsh [1]; the basic idea is also in Fefferman and Stein [1]. See also Taibleson and Weiss [1] for another proof that is closer in spirit to ours.

Characterizations of $(H^p)^*$ $(0 < p \leq 1)$ in terms of BMO and Campanato or Lipschitz spaces have been obtained in various other settings by Calderón and Torchinsky [2], Chang and Fefferman [1], Coifman and Weiss [2], García-Cuerva [1], Garnett and Latter [1], Goldberg [1], and Mauceri, Picardello, and Ricci [1].

The Campanato spaces $C^{\alpha}_{q,N}$ with $1 \leq q < \infty$ were first defined and studied systematically on $\mathbb{R}^n$ by Campanato [1] (with different notation: our $C^{\alpha}_{q,N}$ is Campanato's $\mathcal{L}_N^{(q,\lambda)}$ where $\lambda = q\alpha + n$). Campanato [1] showed that elements of $C^{\alpha}_{q,N}$ satisfy Lipschitz conditions of order α, although his results are local (i.e., they pertain to functions defined on bounded open subsets of $\mathbb{R}^n$) and they do not cover the case $N-\alpha \in \mathbb{N}$. For $\alpha < 1$, $N = 0$, these results were first proved by Meyers [1]; analogous results on spaces of homogeneous type have been obtained by Macías and Segovia [1]. Krantz [2] has proved an abstract generalization of Campanato's theorem.

The Campanato spaces when $q = \infty$ are the simplest versions of these spaces, and so it is ironic that their usefulness has been largely overlooked. (This is probably due to the fact that when $\alpha = 0$ there is an essential difference between $q < \infty$ and $q = \infty$.) Their identification for all $\alpha > 0$ with Lipschitz spaces (locally, for $\mathbb{R}^n$ and several generalizations) seems to appear first in Nagel and Stein [1]; see also Jerison [1].

CHAPTER 6

Convolution Operators on H^p

In this chapter we study the action on H^p of certain types of convolution operators which include the classical singular and fractional integrals. As an application, we prove a Marcinkiewicz-type multiplier theorem for functions of the sub-Laplacian on a stratified group.

A. Kernels of Type (α, r)

The convolution kernels we shall be considering are the following. Suppose $0 < \alpha < Q$ and r is a positive integer. A *kernel of type* (α, r) is a function K on G which is of class $C^{(r)}$ on $G\backslash\{0\}$ and satisfies

$$(6.1) \qquad |Y^I K(x)| \leq A_I |x|^{\alpha-Q-d(I)} \qquad \text{for } |I| \leq r, \quad x \neq 0.$$

In view of Proposition 1.29, it is easily verified that (6.1) is equivalent to

$$(6.1') \qquad |X^I K(x)| \leq A_I' |x|^{\alpha-Q-d(I)} \qquad \text{for } |I| \leq r, \quad x \neq 0.$$

The estimate (6.1) with $I = 0$ implies that K is locally integrable and bounded near infinity, so that K defines a tempered distribution. Moreover, we have:

(6.2) PROPOSITION. *Suppose* $0 < \alpha < Q$, $1 < p < Q/\alpha$, *and* $q^{-1} = p^{-1} - (\alpha/Q)$. *If* K *satisfies* (6.1) *with* $I = 0$, *then the operator* $T_K : f \to f{*}K$ *is bounded from* L^p *to* L^q. *Also,* T_K *is weak type* $(1, Q/(Q-\alpha))$.

Proof: The estimate $|K(x)| \leq A_0|x|^{\alpha-Q}$ implies that $K \in \text{weak } L^{Q/(Q-\alpha)}$, so the result follows from Proposition 1.19. #

More interesting and more subtle is the limiting case $\alpha = 0$. Here the estimate (6.1) does not imply the local integrability of K (and hence that K defines a distribution), nor does it automatically yield an L^p boundedness theorem. Hence we assume these conditions separately, as follows.

If r is a positive integer, a *kernel of type* $(0,r)$ is a distribution K on G which is of class $C^{(r)}$ on $G\backslash\{0\}$ and satisfies (6.1) with $\alpha = 0$ (so that K is necessarily tempered), and which also satisfies

$$(6.3) \qquad \|f*K\|_2 \leq B\|f\|_2 \quad \text{for } f \in \mathcal{S}.$$

(when $G = \mathbb{R}^n$, (6.3) is usually most conveniently expressed by saying that the Fourier transform of K belongs to L^∞.) We shall discuss some examples later, after proving our main theorem. First, a couple of technical lemmas.

(6.4) LEMMA. *Suppose* $0 \leq \alpha < Q$ *and* $r \geq 1$. *There exist* $N \in \mathbb{N}$ *and* $C > 0$ *such that for every kernel* K *of type* (α,r) *and every* $\phi \in \mathcal{S}$,

$$(6.5) \qquad |Y^I K*\phi(x)| \leq C_I\|\phi\|_{(N)}|x|^{\alpha-Q-d(I)},$$

where $C_I \leq C(A_0 + A_I)$ *if* $\alpha > 0$ *and* $C_I \leq C(A_0 + A_I + B)$ *if* $\alpha = 0$, *and* A_I, B *are as in* (6.1) *and* (6.3).

Proof: Fix $\eta \in C_0^\infty(B(1,0))$ such that $\eta = 1$ on $B(1/2,0)$, and set $K^0 = \eta K$, $K^\infty = (1-\eta)K$. Then K^∞ is of class $C^{(r)}$ on G, and

$$|Y^I K^\infty(x)| \leq CA_I(1 + |x|)^{\alpha-Q-d(I)} \quad \text{for } |I| \leq r,$$

where C depends only on the choice of η. Hence if $\phi \in \mathcal{S}$ and $N > Q-\alpha+r$, by Lemma 1.10 we have

(6.6) $$|Y^I K^\infty * \phi(x)| \leq CA_I \|\phi\|_{(N)} \int (1+|xy^{-1}|)^{\alpha-Q-d(I)} (1+|y|)^{-(Q+1)(N+1)} dy$$

$$\leq C'A_I \|\phi\|_{(N)} (1+|x|)^{\alpha-Q-d(I)} \int (1+|y|)^{Q+d(I)-\alpha-(Q+1)(N+1)} d$$

$$\leq C''A_I \|\phi\|_{(N)} |x|^{\alpha-Q-d(I)}.$$

If $\alpha > 0$, fix p with $1 < p < Q/(Q-\alpha)$, and let p' be the conjugate exponent. Since $|K^0(x)| \leq A_0|x|^{\alpha-Q}$ and $\operatorname{supp} K^0 \subset B(1,0)$ we have $K^0 \in L^p$ and $\|K^0\|_p \leq CA_0$, hence

$$|K^0 * \phi(x)| \leq \int |K^0(y)\phi(y^{-1}x)| dy$$

$$\leq CA_0 \left(\int_{|y|\leq 1} |\phi(y^{-1}x)|^{p'} dy\right)^{1/p'}$$

$$\leq C'A_0 \|\phi\|_{(0)} (1+|x|)^{-Q-1}$$

$$\leq C'A_0 \|\phi\|_{(0)} |x|^{\alpha-Q}.$$

To obtain the corresponding result for $Y^I K^0 * \phi$ $(|I| \leq r)$, we use Proposition 1.29 to write $Y^I = \Sigma_{|J|\leq|I|} P_{IJ} X^J$ with $P_{IJ} \in \mathcal{P}_{d(J)-d(I)}$. The above estimate then yields

(6.7) $$|Y^I K^0 * \phi(x)| \leq \Sigma_{|J|\leq|I|} |P_{IJ}(x)| |K^0 * X^J \phi(x)|$$

$$\leq CA_0 (1+|x|)^{\bar{d}r} \|\phi\|_{(N)} (1+|x|)^{-Q-2\bar{d}r}$$

$$\leq CA_0 \|\phi\|_{(N)} |x|^{\alpha-Q-d(I)},$$

provided N is sufficiently large. Combining the estimates (6.6) and (6.7), we have completed the proof for $\alpha > 0$.

The estimate for K^0 in the case $\alpha = 0$ is more difficult. To begin with we observe that $\|K*\psi\|_2 \leq B\|\psi\|_2$ for all $\psi \in \mathcal{S}$, where B is the constant in (6.3). (This follows since the adjoint of the operator $f \to f*K$ is $f \to f*\tilde{K}$, where $\tilde{K}(x) = \overline{K(x^{-1})}$, so $\|f*\tilde{K}\|_2 \leq B\|f\|_2$. But $K*\psi = (\tilde{\psi}*\tilde{K})^{\sim}$.) Hence, by the classical Sobolev inequality (cf. Stein [2], p. 124), if $m > (\dim G)/2$ we have

$$\sup_{|x|\leq 3\gamma} |K*\psi(x)| \leq C\Sigma_{|J|\leq m} \|X^J(K*\psi)\|_2 \leq BC\Sigma_{|J|\leq m} \|X^J\psi\|_2.$$

On the other hand, $K^\infty \in L^2$ and $\|K^\infty\|_2 \leq CA_0$, so

$$\|K^\infty*\psi\|_\infty \leq CA_0\|\psi\|_2.$$

Now suppose $\psi \in C_0^\infty(B(2,0))$. Then $K^0*\psi$ is supported in $B(3\gamma,0)$, so by combining the preceding estimates we obtain

$$(6.8) \qquad \|K^0*\psi\|_\infty \leq C(A_0+B)\Sigma_{|J|\leq m} \|X^J\psi\|_2 \qquad (\psi \in C_0^\infty(B(2,0))).$$

Let us fix $\zeta \in C_0^\infty(B(2,0))$ with $\zeta = 1$ on $B(1,0)$. Given $\phi \in \mathcal{S}$, for each $x \in G$ let $\psi^x(z) = \zeta(z^{-1})\phi(zx)$. Since $\operatorname{supp} K^0 \subset B(1,0)$, by (6.8) we have

$$|K^0*\phi(x)| = \left|\int K^0(y)\zeta(y)\phi(y^{-1}x)dy\right| = \left|\int K^0(y)\psi^x(y^{-1})dy\right| = |K^0*\psi^x(0)|$$
$$\leq \|K^0*\psi^x\|_\infty \leq C(A_0+B)\Sigma_{|J|\leq m} \|X^J\psi^x\|_2.$$

But clearly, if N is sufficiently large,

$$\|X^J\psi^x\|_2 \leq C\|\phi\|_{(N)}(1+|x|)^{\alpha-Q} \qquad (|J| \leq m).$$

This yields the desired estimate for $K^0*\psi$, and the corresponding estimates for $Y^IK^0*\psi$ follow as in the case $\alpha > 0$. #

(6.9) LEMMA. *If* α, r, N, C_I *are as in Lemma* 6.4 *and* K *is a kernel of type* (α, r), *then*

$$M^0_{(N)}(Y^IK)(x) \leq C_I|x|^{\alpha-Q-d(I)} \qquad \text{for } |I| \leq r.$$

Proof: It is easy to check that if K is a kernel of type (α,r) then so is K_s for all $s > 0$. More precisely, if K satisfies (6.1) then K_s also satisfies (6.1) with A_I replaced by $s^{-\alpha}A_I$, and if K satisfies (6.3) then so does K_s (with the same B). Hence K_s satisfies (6.5) with C_I replaced by $s^{-\alpha}C_I$, and we have

$$|Y^IK*\phi_t(x)| = |((Y^IK)_{1/t}*\phi)_t(x)| = t^{-d(I)}|(Y^I(K_{1/t})*\phi)_t(x)|$$

$$\leq t^{-d(I)-Q}\, t^{\alpha}C_I\|\phi\|_{(N)}|x/t|^{\alpha-Q-d(I)}$$

$$= C_I\|\phi\|_{(N)}|x|^{\alpha-Q-d(I)}. \quad \#$$

Now we are ready for the fundamental theorem of this chapter:

(6.10) THEOREM. *Suppose* r *is a positive integer*, $0 \leq \alpha < Q$, $Q/(Q+r) < p < Q/\alpha$, $(1/q) = (1/p) - (\alpha/Q)$, *and* K *is a kernel of type* (α,r). *Then the operator* T_K *defined by* $T_Kf = f*K$ *is bounded from* H^p *to* H^q. *Moreover,* T_K *is weak type* $(1, Q/(Q-\alpha))$.

Proof: We first observe that it suffices to prove boundedness from H^p to H^q for $Q/(Q+r) < p \leq 1$. Indeed, when $\alpha > 0$ the case $p > 1$ and the weak type $(1,Q/(Q-\alpha))$ result are covered by Proposition 6.2. When $\alpha = 0$, the boundedness on H^p, $1 < p < 2$, follows from the boundedness on H^1 and the estimate (6.3) by Theorem 3.34. The result for $2 < p < \infty$ then follows by duality, since the adjoint of T_K is $T_{\tilde{K}}$ where $\tilde{K}(x) = \overline{K(x^{-1})}$. (This is again a kernel of type (α,r), because if K satisfies (6.1) then $\tilde{K}$ automatically satisfies the equivalent condition (6.1').) Finally, the weak type (1,1) result follows by applying Theorem 3.37 to $Tf = M(T_K f)$.

Suppose then that $Q/(Q+r) < p \leq 1$. Fix $\bar{p}$ with $1 < \bar{p} < Q/\alpha$ (in case $\alpha = 0$, take $\bar{p} = 2$), and define $\bar{q}$ by $\bar{q}^{-1} = \bar{p}^{-1} - (\alpha/Q)$. Also, let $a = \max\{a' \in \Delta : a' < r\}$. Then the condition $p > Q/(Q+r)$ means precisely that a is p-admissible. It will therefore suffice to show that if $M = M^0_{(N)}$ where N is as in Lemma 6.4, there exists $C > 0$ such that $\|M(T_K f)\|_q \leq C$ for all $(p,\bar{p},a)$-atoms f.

Let f be a $(p,\bar{p},a)$-atom associated to $B = B(\rho,x_0)$. Without loss of generality we may assume that $x_0 = 0$, and we set $\tilde{B} = B(2\gamma\beta^r\rho,x_0)$. By the maximal theorem (2.4) and Proposition (6.2) (for $\alpha > 0$) or the estimate (6.3) (for $\alpha = 0$),

$$(6.11)\qquad \int_{\tilde{B}} (M(T_K f)(x))^q dx \leq \left[\int_{\tilde{B}} (M(T_K f)(x))^{\bar{q}}\, dx\right]^{q/\bar{q}} |\tilde{B}|^{1-(q/\bar{q})}$$

$$\leq C_1 |B|^{1-(q/\bar{q})} \|M(T_K f)\|_{\bar{q}}^{q}$$

$$\leq C_2 |B|^{1-(q/\bar{q})} \|T_K f\|_{\bar{q}}^{q}$$

$$\leq C_3|B|^{1-(q/\bar{q})} \|f\|_{\bar{q}}^{q}$$

$$\leq C_3|B|^{q[(1/q)-(1/\bar{q})-(1/p)+(1/\bar{p})]}$$

$$= C_3.$$

Next, suppose $\phi \in S$ and $\|\phi\|_{(N)} \leq 1$. For each $x \in \tilde{B}^c$ and $t > 0$ let $P_{t,x}$ be the right Taylor polynomial of $K*\phi_t$ at x of homogeneous degree a. Then by the Taylor inequality (1.37),

$$|K*\phi_t(y^{-1}x) - P_{t,x}(y^{-1})| \leq C_4\Sigma_{|I|\leq r\leq d(I)} |y|^{d(I)} \sup_{|z|\leq\beta^r|y|} |Y^I(K*\phi_t)(zx)|.$$

But by Lemma 6.9,

$$|Y^I(K*\phi_t)(zx)| = |(Y^I K)*\phi_t(zx)| \leq C_5|zx|^{\alpha-Q-d(I)} \qquad (t > 0, \quad |I| \leq r).$$

Also, if $y \in B$, $x \in \tilde{B}^c$, and $|z| \leq \beta^r|y|$ then $|x| \geq 2\gamma|z|$ and hence $|zx| \geq |x|/2$. Thus if $y \in B$,

$$|K*\phi_t(y^{-1}x) - P_{t,x}(y^{-1})| \leq C_6\Sigma_{|I|\leq r\leq d(I)} |y|^{d(I)} |x|^{\alpha-Q-d(I)},$$

and hence, if $\bar{p}'$ is the conjugate exponent to $\bar{p}$,

$$|(T_K f)*\phi_t(x)| = |\int f(y)[K*\phi_t(y^{-1}x) - P_{t,x}(y^{-1})]dy|$$

$$\leq C_6\Sigma_{|I|\leq r\leq d(I)} |x|^{\alpha-Q-d(I)} \|f\|_{\bar{p}} [\int_B |y|^{\bar{p}'d(I)}dy]^{1/\bar{p}'}$$

$$\leq C_7\Sigma_{|I|\leq r\leq d(I)} |x|^{\alpha-Q-d(I)} |B|^{1-(1/p)-(d(I)/Q)}.$$

But

$$\alpha-Q-d(I) \leq \alpha-r-Q < \alpha-Q(p^{-1}-1)-Q = \alpha-(Q/p) = -Q/q,$$

so

$$\int_{\tilde{B}^c} |x|^{q(\alpha-Q-d(I))}dx = C_8|B|^{(q\alpha/Q)-q-(qd(I)/Q)+1}$$

$$= C_8|B|^{q[-1+(1/p)+(d(I)/Q)]}.$$

It follows that

$$\int_{\tilde{B}^c} (M(T_K f)(x))^q dx \leq C_9,$$

and combining this with (6.11) we are done. #

(6.12) Remark: In case G is stratified, we can obtain the same conclusion under somewhat weaker hypotheses on K. Namely, it suffices to assume that $Y^I K$ is continuous on $G\setminus\{0\}$ and satisfies (6.1) merely for $d(I) \leq r$ rather than for $|I| \leq r$. (In case $\alpha = 0$ and $r < \bar{d}$ we must assume in addition that $X_j K$ is continuous on $G\setminus\{0\}$ and satisfies $|X_j K(x)| \leq C|x|^{-Q-1}$ for $j = 1,\ldots,\nu$ (i.e. for $d_j = 1$) in order to obtain the result for $2 < p < \infty$. If $r \geq \bar{d}$ this is implied by the estimate for $Y_j K(x)$ $(j = 1,\ldots,n)$ by Proposition 1.29.) The proof is identical to the one given above except that the stratified Taylor inequality (1.42) is used instead of (1.37) to estimate $M(T_K f)$ on $\tilde{B}^c$.

We now discuss an important class of examples of kernels of type (α,r), namely the homogeneous kernels. If K is a function which is of class $C^{(r)}$ on $G\setminus\{0\}$ and homogeneous of degree $\lambda-Q$ where $0 < \text{Re}\,\lambda < Q$, it is easily verified that K is a kernel of type $(\operatorname{Re}\lambda, r)$. Our object is to show that a similar result holds for $\operatorname{Re}\lambda = 0$. To begin with, we have the following structure theorem for homogeneous distributions of degree $-Q$.

(6.13) PROPOSITION. Let k be a continuous function on $G\setminus\{0\}$ which is homogeneous of degree $-Q$ and satisfies $\mu_k = 0$, where μ_k is defined in Proposition 1.13. Then the formula

$$(6.14) \qquad \langle PV(k), \phi\rangle = \lim_{\varepsilon\to 0}\int_{|x|>\varepsilon} k(x)\phi(x)dx \qquad (\phi \in S)$$

defines a tempered distribution $PV(k)$ which is homogeneous of degree $-Q$. Conversely, suppose K is a tempered distribution which is homogeneous of degree $-Q$ and whose restriction to $G\setminus\{0\}$ is a continuous function k. Then $\mu_k = 0$ and $K = PV(k) + C\delta$ for some $C \in \mathbb{C}$, where δ is the point mass at 0.

Proof: If $\mu_k = 0$, the limit in (6.14) exists for any $\phi \in S$, because

$$\lim_{\varepsilon\to 0}\int_{|x|>\varepsilon} k(x)\phi(x)dx$$

$$= \lim_{\varepsilon\to 0}\int_{\varepsilon<|x|<1} k(x)[\phi(x)-\phi(0)]dx + \int_{|x|\geq 1} k(x)\phi(x)dx$$

$$= \int_{|x|\leq 1} k(x)[\phi(x)-\phi(0)]dx + \int_{|x|\geq 1} k(x)\phi(x)dx,$$

the last integrals being convergent since $|\phi(x)-\phi(0)| = O(|x|)$. From this formula it is clear that $PV(k)$ is continuous on S, and from (6.14) it is easily checked that $\langle PV(k), \phi\circ\delta_r\rangle = \langle PV(k), \phi\rangle$ for any $r > 0$, so that

$PV(k)$ is homogeneous of degree $-Q$ as a distribution. Moreover, if K is a homogeneous distribution which agrees with k away from 0 then $K-PV(k)$ is supported at 0, hence is a linear combination of δ and its derivatives. But $X^I\delta$ is homogeneous of degree $-Q-d(I)$, so by homogeneity we must have $K-PV(k) = C\delta$.

It therefore remains to show that if $k \in C^{(0)}(G\setminus\{0\})$ is the restriction to $G\setminus\{0\}$ of $K \in \mathcal{S}'$ which is homogeneous of degree $-Q$ then $\mu_k = 0$. Consider the distribution F defined by

$$\langle F,\phi\rangle = \int_{|x|\leq 1} k(x)[\phi(x)-\phi(0)]dx + \int_{|x|\geq 1} k(x)\phi(x)dx \qquad (\phi \in \mathcal{S}).$$

(As above, this is well defined since $|\phi(x)-\phi(0)| = O(|x|)$.) F agrees with K away from 0, so again $F-K = \Sigma a_I X^I\delta$. Since K is homogeneous of degree $-Q$, for any $\phi \in \mathcal{S}$ and $r > 0$ we have

$$\langle F,\phi\circ\delta_r\rangle - \langle F,\phi\rangle = \langle F-K,\phi\circ\delta_r\rangle - \langle F-K,\phi\rangle$$

$$= \Sigma a_I(r^{|I|}-1)X^I\phi(0),$$

which is bounded as $r \to 0$. However, for $r < 1$,

$$\langle F,\phi\circ\delta_r\rangle - \langle F,\phi\rangle = \phi(0)\int_{r<|x|<1} k(x)dx = -\phi(0)\mu_k \log r,$$

which is a contradiction when $\phi(0) \neq 0$ unless $\mu_k = 0$. #

A similar result holds for kernels which are homogeneous of degree $\lambda-Q$ where $\lambda \neq 0$ and $\mathrm{Re}\,\lambda = 0$:

(6.15) Proposition. *Let* k *be a continuous function on* $G\backslash\{0\}$ *which is homogeneous of degree* $\lambda-Q$, *where* $\lambda \neq 0$ *and* $\operatorname{Re}\lambda = 0$; *let* $k'(x) = k(x)|x|^{-\lambda}$ *and define the mean-value* $\mu_{k'}$ *as in* (1.13). *Set* $R = e^{2\pi/|\lambda|}$. *Then for any* $j \in \mathbb{Z}$

$$\int_{R^j \leq |x| \leq R^{j+1}} k(x)dx = 0 \tag{6.16}$$

and the formula

$$\langle PV(k),\phi\rangle = \lim_{j\to-\infty} \int_{|x|\geq R^j} k(x)\phi(x)dx + \mu_{k'}\phi(0)/\lambda \qquad (\phi \in S) \tag{6.17}$$

defines a tempered distribution $PV(k)$ *which is homogeneous of degree* $\lambda-Q$. *Moreover it is the only distribution which is homogeneous of degree* $\lambda-Q$ *and agrees with* k *on* $G\backslash\{0\}$.

Proof: By Proposition 1.13 we have (since $k(x) = k'(x)|x|^{\lambda}$)

$$\int_{R^j \leq |x| \leq R^{j+1}} k(x)dx = \mu_{k'}[e^{2\pi(j+1)\lambda/|\lambda|} - e^{2\pi j\lambda/|\lambda|}]/\lambda,$$

which vanishes since $\lambda/|\lambda| = \pm i$. This establishes (6.16), and from this the existence of the limit in (6.17) follows as in Proposition (6.13). The argument also shows that for any $\delta > 0$

$$\langle PV(k),\phi\rangle = \int_{|x|\leq\delta} k(x)[\phi(x)-\phi(0)]dx + \int_{|x|\geq\delta} k(x)\phi(x)dx + \mu_{k'}\delta^{\lambda}\phi(0)/\lambda$$

and hence the homogeneity of $PV(k)$ is an easy consequence. The uniqueness is true because there are no distributions supported at the origin which are homogeneous of degree $\lambda-Q$. #

We now have some candidates for kernels of type $(0,r)$. It remains to verify the L^2 boundedness condition (6.3). For this purpose we employ the following lemma:

(6.18) LEMMA. Let $\phi : \mathbb{Z} \to (0,\infty)$ satisfy $\Sigma_{-\infty}^{\infty} \sqrt{\phi(j)} = A < \infty$, and let $\mathcal{X}$ and $\mathcal{Y}$ be Hilbert spaces. If $\{T_j\}_{-\infty}^{\infty}$ is a sequence of bounded linear maps from $\mathcal{X}$ to $\mathcal{Y}$ such that $\|T_i^*T_j\| \leq \phi(j-i)$ and $\|T_iT_j^*\| \leq \phi(j-i)$ for all $i,j \in \mathbb{Z}$, then $\|\Sigma_M^N T_j\| \leq A$ for all $M,N \in \mathbb{Z}$.

Proof: Let $T = \Sigma_M^N T_j$. We have $\|T_j\|^2 = \|T_j^*T_j\| \leq \phi(0) \leq A^2$, and also $\|T\|^2 = \|T^*T\|$. Since T^*T is a self-adjoint operator on $\mathcal{X}$, its norm is its spectral radius:

$$\|T^*T\| = \limsup_{n\to\infty} \|(T^*T)^n\|^{1/n}.$$

But

$$(T^*T)^n = \Sigma_{i_1,i_2,\ldots,i_{2n}} T_{i_1}^*T_{i_2}T_{i_3}^*T_{i_4}\cdots T_{i_{2n-1}}^*T_{i_{2n}}$$

where each index i_j runs from M to N. Now

$$\|T_{i_1}^*T_{i_2}\cdots T_{i_{2n-1}}^*T_{i_{2n}}\| \leq \phi(i_2-i_1)\phi(i_4-i_3)\cdots\phi(i_{2n}-i_{2n-1})$$

and also

$$\|T_{i_1}^*T_{i_2}\cdots T_{i_{2n-1}}^*T_{i_{2n}}\| \leq A\phi(i_3-i_2)\phi(i_5-i_4)\cdots\phi(i_{2n-1}-i_{2n-2})A.$$

Taking the geometric mean of these inequalities,

$$\|T_{i_1}^*T_{i_2}\cdots T_{i_{2n-1}}^*T_{i_{2n}}\| \leq A[\phi(i_2-i_1)\phi(i_3-i_2)\cdots\phi(i_{2n}-i_{2n-1})]^{1/2}.$$

Sum first on i_{2n}, then on i_{2n-1}, etc., up to i_2, obtaining

$$\|(T^*T)^n\| \leq A\Sigma_{i_1=M}^{N} A^{2n-1} = (N-M+1)A^{2n}.$$

Therefore $\|T\|^2 \leq \lim_{n\to\infty} (N-M+1)^{1/n}A^2 = A^2$. #

(6.19) THEOREM. Suppose $\mathrm{Re}\,\lambda = 0$ and r is a positive integer. If $K \in S'$ is homogeneous of degree $\lambda - Q$ and of class $C^{(r)}$ on $G\backslash\{0\}$, then K satisfies (6.3) and hence is a kernel of type $(0,r)$.

Proof: We deal first with the case $\lambda = 0$. In the terminology of Proposition 6.13, we then have $K = PV(k) + C\delta$, and it clearly suffices to consider the case $C = 0$. For $j \in \mathbb{Z}$, set $k_j(x) = k(x)$ if $2^j \leq |x| \leq 2^{j+1}$ and $k_j(x) = 0$ otherwise: thus $k = \Sigma_{-\infty}^{\infty} k_j$ pointwise on $G\backslash\{0\}$ and $K = \Sigma_{-\infty}^{\infty} k_j$ in the sense of distributions. For $f \in L^2$ we define $T_jf = f*k_j$ (which is in L^2 since $k_j \in L^1$), and we set $T_M^N = \Sigma_M^N T_j$.

If $f \in C_0^\infty$ then $T_M^N f \to f*K$ in the distribution topology as $M \to -\infty$, $N \to +\infty$, and we claim that this convergence also takes place in the L^2 norm. Indeed, if $M < 0 < N$ we write $T_M^N = T_M^0 + T_1^N$. Since k is square-integrable on $U = \{x : |x| \geq 2\}$, $\Sigma_1^N k_j$ converges in L^2 as $N \to \infty$ to $k\chi_U$, so $T_1^N f \to f*(k\chi_U)$ in the L^2 norm because $f \in L^1$. On the other hand, since $\mu_k = 0$,

$$\lim_{M\to-\infty} T_M^0 f(x) = \int_{|y|\leq 1} [f(xy^{-1}) - f(x)]k(y)dy,$$

the convergence being uniform in x. Since $T_M^0 f$ is supported in the compact set $\{xy : x \in \mathrm{supp}\, f, |y| \leq 2\}$, the convergence is also in L^2.

In order to prove (6.3), it therefore remains to show that the operators T_M^N are uniformly bounded on L^2. First, we observe that

$$\|T_j\| \leq \|k_j\|_1 = \int_{2^j \leq |y| < 2^{j+1}} |k(y)|dy \leq C_0 \int_{2^j \leq |y| < 2^{j+1}} |y|^{-Q}dy = C.$$

Hence $\|T_i^*T_j\| \leq C^2$ and $\|T_iT_j^*\| \leq C^2$ for all i,j, so by Lemma 6.18 it will suffice to show that

$$\|T_i^*T_j\| \leq C2^{-|i-j|} \quad \text{and} \quad \|T_iT_j^*\| \leq C2^{-|i-j|}$$

whenever $|i-j|$ is sufficiently large; in fact, we shall obtain these estimates for $|i-j| \geq 3 + \log_2 \gamma$.

We observe that

$$T_j^*f(x) = \int f(xz^{-1})\overline{k_j(z^{-1})}dz,$$

and hence

$$T_iT_j^*f(x) = \iint f(xy^{-1}z^{-1})\overline{k_j(z^{-1})}k_i(y)dzdy$$

$$= f*G_{ij}(x)$$

where

$$G_{ij}(z) = \int k_i(yz)\overline{k_j(y)}dy.$$

We can then estimate $\|T_iT_j^*\|$ by estimating $\|G_{ij}\|_1$. Also, $T_i^*T_j$ is of the same form except that $k(x)$ is replaced by $\overline{k(x^{-1})}$, so the same arguments will yield the same estimates for $\|T_i^*T_j\|$. Moreover,

$T_jT_i^*f = (T_iT_j^*)^*f = f*\tilde{G}_{ij}$ where $\tilde{G}_{ij}(z) = \overline{G_{ij}(z^{-1})}$, so we can interchange i and j. In short, we are reduced to proving that

$$\|G_{ij}\|_1 \leq C2^{j-i} \quad \text{when} \quad i-j \geq 3 + \log_2 \gamma.$$

Given such an i and j, let

$$E = \{(y,z) : 2^i \leq |yz| < 2^{i+1},\ 2^j \leq |y| < 2^{j+1}\}, \quad E_z = \{y : (y,z) \in E\}.$$

Then since $\mu_k = 0$,

$$G_{ij}(z) = \int_{E_z} k_i(yz)\overline{k_j(y)}dy - k_i(z) \int k_j(y)dy$$

$$= \int_{E_z} [k_i(yz) - k_i(z)]\overline{k_j(y)}dy - k_i(z) \int_{y \notin E_z} k_j(y)dy$$

$$= F_{ij}(z) + H_{ij}(z).$$

Since $k_i(yz) = k(yz)$ for $(y,z) \in E$,

$$\int |F_{ij}(z)|dz = \iint_E |k(yz) - k(z)|\,|k_j(y)|dydz + \iint_E |k(z) - k_i(z)|\,|k_j(y)|dydz$$

$$= I_1 + I_2.$$

If $(y,z) \in E$ we have

$$|z| \leq \gamma(|yz| + |y|) \leq \gamma(2^{i+1} + 2^{j+1}) \leq \gamma 2^{i+2}.$$

Also, $|yz| \leq \gamma(|y| + |z|)$ and $\gamma \leq 2^{i-j-3}$, so

$$|z| \geq \gamma^{-1}|yz| - |y| \geq 2^{3+j-i}\cdot 2^i - 2^{j+1} \geq 2^{j+2},$$

and in particular $|z| \geq 2|y|$. Thus by Proposition 1.7,

$$I_1 \leq \sup_E |k(yz) - k(z)| \int |k_j(y)|\,dy \int_{|z| \leq \gamma 2^{i+2}} dz$$

$$\leq C' \sup_E |y||z|^{-Q-1}(\gamma 2^{i+2})^Q$$

$$\leq C' 2^{j+1} 2^{i(-Q-1)}(\gamma 2^{i+2})^Q$$

$$= C'' 2^{j-i}.$$

In I_2 the integrand vanishes unless $|z| < 2^i$ or $|z| \geq 2^{i+1}$, in which case $|k_i(z) - k(z)| = |k(z)|$. Also, for $|z| < 2^i$ and $(y,z) \in E$,

$$2^i - |z| \leq |yz| - |z| \leq \gamma |y| \leq \gamma 2^{j+1},$$

so that

$$|z| \geq 2^i(1 - \gamma 2^{j-i+1}).$$

(Our condition on $i-j$ ensures that $1 - \gamma 2^{j-i+1} \geq 3/4$.) Likewise, if $|z| \geq 2^{i+1}$ and $(y,z) \in E$,

$$|z| - 2^{i+1} \leq |z| - |yz| \leq \gamma |y| \leq \gamma 2^{j+1},$$

so that

$$|z| \leq 2^{i+1}(1 + \gamma 2^{j-i}).$$

Hence I_2 is dominated by

$$\int |k_j(y)|dy\Big[\Big(\int_{2^i(1-\gamma 2^{j-i+1})\le|z|\le 2^i} + \int_{2^{i+1}\le|z|\le 2^{i+1}(1+\gamma 2^{j-i})}\Big)|z|^{-Q}dz\Big]$$

$$\le C[|\log(1-\gamma 2^{j-i+1})| + \log(1+\gamma 2^{j-i})]$$

$$\le C' 2^{j-i}.$$

Therefore,

$$\int |F_{ij}(z)|dz \le C2^{j-i}.$$

On the other hand,

$$\int |H_{ij}(z)|dz = \iint_{E^c} |k_i(z)||k_j(y)|dydz$$

The integrand vanishes unless $2^i \le |z| \le 2^{i+1}$ and $2^j \le |y| \le 2^{j+1}$, in which case either $|yz| \ge 2^{i+1}$ or $|yz| < 2^i$ since $(y,z) \notin E$. As above, then, if $|yz| < 2^i$,

$$|z| - 2^i \le |z| - |yz| \le \gamma|y| \le \gamma 2^{j+1}, \quad \text{so} \quad |z| \le 2^i(1+\gamma 2^{j-i+1}),$$

while if $|yz| \ge 2^{i+1}$,

$$2^{i+1} - |z| \le |yz| - |z| \le \gamma|y| \le \gamma 2^{j+1}, \quad \text{so} \quad |z| \ge 2^{i+1}(1-\gamma 2^{j-i}).$$

Hence, as in the estimate for I_2,

$$\int |H_{ij}(z)|dz \le C2^{j-i},$$

which completes the proof for $\lambda = 0$.

The proof for $\lambda \neq 0$ is exactly the same, except that we replace the annuli $2^j \leq |x| < 2^{j+1}$ by the annuli $R^j \leq |x| < R^{j+1}$ where $R = e^{2\pi/|\lambda|}$, and use Proposition 6.15 and the equation (6.16) instead of Proposition 6.13 and the equation $\mu_K = 0$. #

These results can be generalized to vector-valued functions. As in the concluding remarks of Chapter 3, Section B, if X is any Banach space we denote spaces of functions and distributions on G with values in X by appending a subscript X: thus, L^p_X, S'_X, etc. Suppose X and Y are Banach spaces, and let $B(X,Y)$ be the space of bounded linear maps from X to Y. We define $B(X,Y)$-valued kernels of type (α,r) just as in the scalar case, except that $|Y^I K(x)|$ is to be replaced by $\|Y^I K(x)\|_{B(X,Y)}$ in (6.1), and (6.3) is to be replaced by

$$\|f*K\|_{L^2_Y} \leq B\|f\|_{L^2_X} \qquad (f \in S_X).$$

We then have the following generalization of Theorems 6.10 and 6.19:

(6.20) THEOREM. (a) <u>Suppose</u> r <u>is a positive interger</u>, $0 \leq \alpha < Q$, $Q/(Q+r) < p < Q/\alpha$, $(1/q) = (1/p) - (\alpha/Q)$, <u>and</u> X <u>and</u> Y <u>are Banach spaces</u>. <u>If</u> K <u>is a</u> $B(X,Y)$-<u>valued kernel of type</u> (α,r), <u>then the operator</u> $f \to f*K$ <u>is bounded from</u> H^p_X to H^q_Y.

(b) <u>Suppose</u> $\mathrm{Re}\,\lambda = 0$, r <u>is a positive integer, and</u> X <u>and</u> Y are <u>Hilbert spaces</u>. <u>If</u> $K \in S'_{B(X,Y)}$ <u>is homogeneous of degree</u> $\lambda - Q$ <u>and of class</u> $C^{(r)}$ <u>on</u> $G\backslash\{0\}$, <u>then</u> K <u>is a</u> $B(X,Y)$-<u>valued kernel of type</u> $(0,r)$.

Proof: In view of the remarks at the end of Chapter 3, Section B, the proof of (a) is the same as the proof of Theorem 6.10, with absolute values replaced by norms in appropriate spots. If X and Y are Hilbert spaces, then so are L^2_X and L^2_Y, so Lemma 6.18 still applies. The proof of Theorem 6.19, with complex conjugates replaced by adjoints, then establishes (b). #

Further results concerning kernels of type (α,r) can be obtained by applying the duality theorems of Chapter 5. (For simplicity, we restrict attention to the scalar case.) The pairing between the H^p spaces and the Campanato spaces is given by $(f,u) \to \int fu$, at least for f in a suitable dense subspace of H^p. Thus, if K is a kernel of type (α,r), the dual of the operator $T_K : f \to f*K$ is, at least formally, $T_{\tilde{K}}$, where $\tilde{K}(x) = K(x^{-1})$. (We say "formally" because the integrals defining $T_{\tilde{K}}u = u*\tilde{K}$ are likely to be divergent, so one must be careful in interpreting $T_{\tilde{K}}$.) Since $\tilde{K}$ is a kernel of type (α,r) whenever K is (by the equivalence of (6.1) and (6.1')), we see that — again, formally — kernels of type (α,r) define bounded operators from $(H^p)^*$ to $(H^q)^*$ for appropriate p and q.

These results as applied to Campanato spaces $C^\alpha_{q,a}$ with $\alpha > 0$ allow one to show that convolution with a kernel of type (α_0,r) increases the local Lipschitz smoothness by the amount α_0 (in the appropriate sense). This can be deduced by applying the duality and using the observation that if q is a fixed C^∞ function of compact support, then the mapping $f \to f\cdot\phi$ is continuous from $C^\alpha_{q,a} \cap L^\infty$ to $C^\alpha_{q,a}$. The local regularity results, however, can usually be obtained by more elementary direct arguments: see Folland [1], Korányi and Vági [1], Nagel and Stein [1], and Rothschild and Stein [1]. Here, we wish to consider BMO in more detail.

Suppose $K \in S'$ is of class $C^{(1)}$ on $G\setminus\{0\}$ and homogeneous of degree $\lambda-Q$ with $\operatorname{Re}\lambda = 0$. Then by Theorems 6.10 and 6.19 and Corollary 5.7, K defines a bounded operator T_K on BMO/P_0, namely the dual of $T_{\tilde{K}}$ on H^1. We shall now describe this operator explicitly. Let $K^0(x) = K(x)$ if $|x| \leq 1$ and $K^0(x) = 0$ otherwise, and let $K^\infty = K - K^0$. The proof of Theorem 6.19 shows that the operator $f \to f*K^0$ is bounded on L^2. However, since K^0 has compact support, it follows that $f*K^0$ is well defined as a locally L^2 function whenever f is locally in L^2, and in particular (by Corollary 5.8) whenever $f \in BMO$.

Suppose then that $f \in BMO$ and g is a finite linear combination of $(1,\infty,0)$-atoms. We clearly have

$$\int (f*K^0)(x)g(x)dx = \int f(x)(g*\tilde{K}^0)(x)dx,$$

so that T_{K^0} is the dual of $T_{\tilde{K}^0}$. On the other hand, since $\int g = 0$,

$$g*\tilde{K}^\infty(x) = \int K^\infty(x^{-1}y)g(y)dy = \int [K^\infty(x^{-1}y) - K^\infty(x^{-1})]g(y)dy.$$

Moreover, by Proposition 1.7 we have

$$|K^\infty(x^{-1}y) - K^\infty(x^{-1})| = O(|x|^{-Q-1}) \quad \text{as} \quad x \to \infty,$$

so by Proposition 5.9, the integral

$$\int f(x)[K^\infty(x^{-1}y) - K^\infty(x^{-1})]dx$$

converges for every y and defines a function which we shall call $T_{K^\infty}f$, and we have

$$\int f(x)(g*\tilde{K}^\infty)(x)dx = \int T_{K^\infty} f(y)g(y)dy.$$

Thus T_{K^∞} is the dual of $T_{\tilde{K}^\infty}$, so the dual of $T_{\tilde{K}} : H^1 \to H^1$ is $T_K = T_{K^0} + T_{K^\infty}$. Of course, if $f \in BMO \cap L^p$ for some $p \in (1,\infty)$, our present definition of $T_K f$ differs from $f*K \in L^p$ by the additive constant $\int f(x)K^\infty(x^{-1})dx$ (which is finite since $K^\infty \in L^q$ for $q > 1$). But this is immaterial because T_K is only supposed to act on BMO/P_0, i.e., BMO modulo constants.

Next, suppose K is a kernel of type $(\alpha,1)$, $0 < \alpha < Q$. Then our theorems imply that K defines a bounded operator T_K from $L^{Q/\alpha}$ to BMO. The convolution $f*K$ need not converge for arbitrary $f \in L^{Q/\alpha}$, but it does so (almost everywhere) when f has compact support, so we can define $T_K f$ in general by a limiting process. We now present a direct proof of a slightly stronger result.

(6.21) THEOREM. _If_ K _is a kernel of type_ $(\alpha,1)$, $0 < \alpha < Q$, _then_ T_K _is bounded from_ weak $L^{Q/\alpha}$ _to_ BMO.

Proof: For simplicity we write T instead of T_K, and we employ the terminology of distribution functions and nonincreasing rearrangements introduced in Chapter 1, Section A. If suffices to show that there is a constant C such that $\|Tf\|_{BMO} \leq C$ for all compactly supported $f \in$ weak $L^{Q/\alpha}$ such that $[f]_{Q/\alpha} = 1$. Given such an f and a ball B, we wish to study the behavior of Tf on B. By translation invariance we may assume that $B = B(R,0)$, and we set $\tilde{B} = B(2\gamma\beta R,0)$. Let $f' = f\chi_{\tilde{B}}$ and $f'' = f-f' = f\chi_{\tilde{B}^c}$. Fix p such that $1 < p < Q/\alpha$ and define q by $q^{-1} = p^{-1} - (\alpha/Q)$. Then by Proposition 6.2,

$$\int_B |Tf'(x)|^q dx \leq \int |Tf'(x)|^q dx \leq C_1 (\int |f'(x)|^p dx)^{q/p}$$

$$= C_1 (\int_0^\infty f'*(t)^p dt)^{q/p} \leq (\int_0^{|\tilde{B}|} t^{-\alpha p/Q} \, dt)^{q/p}$$

$$= C_2 |\tilde{B}|^{q[1-(\alpha p/Q)]/p}$$

$$= C_2 |\tilde{B}| .$$

Hence by Hölder's inequality,

$$(6.22) \qquad \int_B |Tf'(x)| dx \leq C_2^{1/q} \, |\tilde{B}| = C_2^{1/q} (2\gamma\beta)^Q \, |B| .$$

Next, let

$$a_B = \int K(y^{-1}) f''(y) dy .$$

(The integral converges since supp f'' is compact and disjoint from the origin.) Then

$$Tf''(x) - a_B = \int_{\tilde{B}^c} f(y)[K(y^{-1}x) - K(y^{-1})]dy .$$

But if $x \in B$ and $y \in B^c$, $|z| \leq \beta|x|$ implies $|y^{-1}z| \geq |y|/2$, so by (6.1) and the left-invariant version of the mean value theorem (1.33),

$$|K(y^{-1}x) - K(y^{-1})| \leq C_3 \Sigma_1^n \, |x|^{d_j} \, |y|^{\alpha - Q - d_j}$$

$$\leq C_3 \Sigma_1^n \, |B|^{d_j/Q} \, |y|^{\alpha - Q - d_j} \qquad (x \in B, \ y \in \tilde{B}^c).$$

Hence,

$$(6.23)\qquad |Tf''(x) - a_B| \le C_3\Sigma_1^n\, |B|^{d_j/Q} \int_{\tilde{B}^c} |f(y)|\,|y|^{\alpha-Q-d_j}\, dy \qquad (x \in B).$$

Next, let $A = |B|^{-\alpha/Q}$, and define $f_1(x) = f(x)$ if $|f(x)| > A$, $f_1(x) = 0$ otherwise, and $f_2 = f - f_1$. Also, fix r such that $Q/\alpha < r < Q/(\alpha-1)$ if $\alpha \ge 1$ and $Q/\alpha < r \le \infty$ if $\alpha < 1$. Then if r' is the conjugate exponent to r, it follows that $(\alpha-Q-d_j)r' < -Q$ for all j, and we have

$$\int_{\tilde{B}^c} |f(y)|\,|y|^{\alpha-Q-d_j}\, dy \le \int_{\tilde{B}^c} |f_1(y)|\,|y|^{\alpha-Q-d_j}\, dy + \int_{\tilde{B}^c} |f_2(y)|\,|y|^{\alpha-Q-d_j}\, dy$$

$$\le |\tilde{B}|^{(\alpha-Q-d_j)/Q}\|f_1\|_1 + \Big(\int_{\tilde{B}^c} |y|^{(\alpha-Q-d_j)r'} dy\Big)^{1/r'} \Big(\int |f_2(y)|^{r'} dy\Big)^{1/r}$$

$$\le C_4\big[\,|B|^{(\alpha-Q-d_j)/Q}\|f_1\|_1 + |B|^{(1/r')+(\alpha-Q-d_j)/Q}\|f_2\|_r\big].$$

But $f_1^*(t) = f^*(t)$ if $t \le \lambda_f(A)$ and $f_1^*(t) = 0$ otherwise, so since $\lambda_f(A) \le A^{-Q/\alpha} = |B|$,

$$\|f_1\|_1 = \|f_1^*\|_1 = \int_0^{\lambda_f(A)} f^*(t)\,dt \le \int_0^{\lambda_f(A)} t^{-\alpha/Q}\, dt$$

$$= C_5\lambda_f(A)^{1-(\alpha/Q)} \le C_5|B|^{1-(\alpha/Q)}.$$

Also, $f_2^*(t) = f^*(t+\lambda_f(A))$ and $f_2^*(t) \le \|f_2\|_\infty \le A$, so

$$\|f_2\|_r = \|f_2^*\|_r = [\int_{\lambda_f(A)}^{\infty} f^*(t)^r dt]^{1/r}$$

$$\leq [\int_{\lambda_f(A)}^{A^{-Q/\alpha}} A^r dt + \int_{A^{-Q/\alpha}}^{\infty} t^{-\alpha r/Q}\, dt]^{1/r}$$

$$\leq C_6[A^r \cdot A^{-Q/\alpha} + A^{-(Q/\alpha)[1-(\alpha r/Q)]}]^{1/r}$$

$$= C_7|B|^{(1/r)-(\alpha/Q)}.$$

Therefore

$$\int_{\tilde{B}^c} |f(y)||y|^{\alpha-Q-1}\, dy$$

$$\leq C_4[C_5|B|^{(\alpha-Q-d_j)/Q}\, |B|^{1-(\alpha/Q)} + C_7|B|^{(1/r')+(\alpha-Q-d_j)/Q}|B|^{(1/r)-(\alpha/Q)}]$$

$$= C_8|B|^{-d_j/Q},$$

and so by (6.23), we have

$$|Tf''(x) - a_B| \leq nC_3C_8 \qquad \text{for} \quad x \in B. \tag{6.24}$$

Finally, combining (6.22) and (6.24), we obtain

$$|B|^{-1} \int_B |Tf(x) - a_B|\, dx \leq C_9.$$

The theorem now follows immediately from Proposition 5.6. #

B. A Multiplier Theorem

In this section we assume that G is a stratified group. We recall from Chapter 1, Section G that $\mathcal{L}$, $h(x,t)$, and $\{H_t\}_{t>0}$ are the sub-Laplacian, heat kernel, and heat semigroup on G. $\{H_t\}_{t>0}$ is a self-adjoint contraction semigroup on L^2, so it has a spectral resolution:

$$H_t = \int_0^\infty e^{-\lambda t}\, dP(\lambda).$$

(The integration is over the open interval $(0,\infty)$, for if $f\in L^2$ and $H_t f = f$ for all t, we have $\|f\|_\infty \le \|f\|_2 \|h(\cdot,t)\|_2 \to 0$ as $t\to\infty$, hence $f = 0$.) If M is a bounded Borel function on $(0,\infty)$, we define the bounded operator $M(\mathcal{L})$ on L^2 by

$$M(\mathcal{L}) = \int_0^\infty M(\lambda)dP(\lambda)$$

Our aim is to prove the following Marcinkiewicz-type multiplier theorem for the operators $M(\mathcal{L})$:

(6.25) THEOREM. <u>Suppose</u> M <u>is of class</u> $C^{(s)}$ <u>on</u> $(0,\infty)$ <u>and</u>

$$\sup_{\lambda>0} |\lambda^j M^{(j)}(\lambda)| \le C < \infty \quad \underline{\text{for}} \quad 0 \le j \le s.$$

<u>If</u> r <u>is a positive integer and</u> $s > r + (3Q/2) + 2$, <u>then</u> $M(\mathcal{L})$ <u>is bounded on</u> H^p <u>for</u> $Q/(Q+r) < p < \infty$.

The idea of the proof is as follows. By the Schwartz kernel theorem (cf. Trèves [1], Chapter 51), $M(\mathcal{L})$ has a tempered distribution kernel; that is, there exists $K\in\mathcal{S}'$ such that $M(\mathcal{L})f = f*K$ for $f\in\mathcal{S}$. We shall show, essentially, that K is a kernel of type $(0,r)$. To accomplish this, we shall require several lemmas. First, let $B^1 = \overline{B(1,0)}$, $B^j = \{x_1 x_2 \cdots x_j : x_i \in B^1 \text{ for } 1 \le i \le j\}$, and $\nu(x) = \inf\{j : x \in B^j\}$.

(6.26) LEMMA. $\nu(xy) \le \nu(x) + \nu(y)$, _and there exists_ $\eta > 0$ _such that_ $\nu(x) \ge \eta|x|$ _for all_ $x \in G$.

Proof: The first assertion is obvious, since if $x \in B^i$ and $y \in B^j$ then $xy \in B^{i+j}$. The second assertion is a theorem of Jenkins [1]. #

Next, let $\mu(x) = e^{\nu(x)}$. Thus by Lemma 6.26 we have

$$\mu(xy) \le \mu(x)\mu(y), \qquad \mu(x) \ge e^{\eta|x|}.$$

If f is a measurable function on G, we define

$$\|f\|_\mu = \int |f(x)|\mu(x)dx.$$

(6.27) LEMMA. $\|f*g\|_\mu \le \|f\|_\mu \|g\|_\mu$.

Proof: $\|f*g\|_\mu \le \iint |f(y)g(y^{-1}x)|\mu(y)dydx$

$$\le \iint |f(y)g(y^{-1}x)|\mu(y)\mu(y^{-1}x)dydx = \|f\|_\mu \|g\|_\mu. \quad \#$$

The next lemma is a theorem of Hulanicki [1], [2]:

(6.28) LEMMA. $\|h(\cdot,t)\|_\mu < \infty$ _for all_ $t > 0$.

(6.29) LEMMA. _If_ M _is a bounded Borel function on_ $(0,\infty)$, _let_ K _be the distribution kernel of_ $M(L)$. _Then for any_ $t > 0$, _if_ $M_{(t)}(\lambda) = M(t\lambda)$, _the distribution kernel of_ $M_{(t)}(L)$ _is_ $K_{\sqrt{t}}$.

Proof: The homogeneity of L means that $[L(f\circ\delta_r)]\circ\delta_{1/r} = r^2 Lf$, so that

$$M_{(r^2)}(L)(f)(x) = M(r^2 L)(f)(x) = [M(L)(f\circ\delta_r)](x/r)$$

$$= \int f(ry)K(y^{-1}(x/r))dy = f*K_r(x).$$

The result follows by setting $t = r^2$. #

Next, for any $m\in\mathbb{Z}$, let

$$E_m(x) = \Sigma_1^\infty \frac{(im)^k h(x,k)}{k!}$$

Since $h(x,k)$ is the convolution of $h(x,1)$ with itself k times, and $h(x,1)$ is the distribution kernel of $\exp(-L)$, it follows that E_m is the distribution kernel of $M_m(L)$ where $M_m(\lambda) = \exp(ime^{-\lambda}) - 1$.

(6.30) LEMMA. $E_m\in L^1$, <u>and for each</u> $\alpha\geq 0$ <u>there exists</u> $C_\alpha > 0$ <u>such that for all</u> $m\in\mathbb{Z}$,

$$\int |x|^\alpha |E_m(x)|dx \leq C_\alpha |m|^{\alpha+(Q/2)+1}.$$

Proof: First, by Lemmas 6.27 and 6.28,

$$\int |E_m(x)|\mu(x)dx \leq \Sigma_1^\infty \frac{|m|^k}{k!}\|h(\cdot,1)\|_\mu^k \leq e^{\theta|m|}, \tag{6.31}$$

where $\theta = \|h(\cdot,1)\|_\mu$. In particular, $E_m\in L^1$. Next, observe that

$$E_m = M(L)h(\cdot,1), \quad \text{where } M(\lambda) = [\exp(ime^{-\lambda}) - 1]/e^{-\lambda}$$

(this is clear by expanding the exponential in a power series), so by the spectral theorem,

$$(6.32)\qquad \int |E_m(x)|^2 dx \le \sup_{\lambda>0} |(\exp(ime^{-\lambda}) - 1)/e^{-\lambda}|^2 \int h(x,1)^2 dx$$

$$\le Cm^2.$$

To estimate $\int |x|^\alpha |E_m(x)| dx$ we estimate separately the integrals over the regions $|x| \le A$ and $|x| > A$, where A will be determined below. By (6.32),

$$(6.33)\qquad \int_{|x|\le A} |x|^\alpha |E_m(x)| dx \le (\int_{|x|\le A} |x|^{2\alpha} dx)^{1/2} (\int |E_m(x)|^2 dx)^{1/2}$$

$$\le C' A^{\alpha+(Q/2)} |m|.$$

On the other hand, by (6.31) and Lemma 6.26,

$$(6.34)\qquad \int_{|x|\ge A} |x|^\alpha |E_m(x)| dx \le \sup_{|x|\ge A} [|x|^\alpha/\mu(x)] \int |E_m(x)| \mu(x) dx$$

$$\le e^{\theta|m|} \sup_{|x|\ge A} |x|^\alpha e^{-\eta|x|}.$$

The estimate we are trying to prove is trivial for small m by (6.31), since $|x|^\alpha \le C_\alpha \mu(x)$, so we may assume that $|m| \ge \alpha/\theta$. We then choose $A = \theta|m|/\eta$, so that $A \ge \alpha/\eta$ and hence

$$\sup_{|x|\ge A} |x|^\alpha e^{-\eta|x|} = A^\alpha e^{-\eta A}.$$

Then by (6.33) and (6.34),

$$\int |x|^\alpha |E_m(x)| dx \le C'(\theta|m|/\eta)^{\alpha+(Q/2)} |m| + e^{\theta|m|}(\theta|m|/\eta)^\alpha e^{-\theta|m|}$$

$$\le C_\alpha |m|^{\alpha+(Q/2)+1}. \quad \#$$

With these preliminaries out of the way, we can get down to business. If M is a function of class $C^{(s)}$ on $(0,\infty)$, we define

$$\|M\|_{C^{(s)}} = \Sigma_0^s \|M^{(j)}\|_\infty.$$

(6.35) LEMMA. Suppose $\alpha \geq 0$ and M is a function of class $C^{(s)}$ supported in $[1/2,2]$, where $s > \alpha + (Q/2) + 2$. Let K be the distribution kernel of $M(L)$. Then $K \in L^1$, and

$$\int |x|^\alpha |K(x)|\,dx \leq C_{s,\alpha} \|M\|_{C^{(s)}}.$$

Proof: Let $F(t) = M(-\log t)$ for $0 < t < 1$ and $F(t) = 0$ otherwise, so that F is supported in $[e^{-2},e^{-1/2}]$ and the $C^{(s)}$ norms of F and M are comparable. Expand F in a Fourier series on $(-\pi,\pi)$: $F(t) = \Sigma_{-\infty}^\infty a_m e^{imt}$. Then $\Sigma a_m = F(0) = 0$, and $|a_m| \leq \|F\|_{C^{(s)}}(1+|m|)^{-s}$. But

$$M(\lambda) = F(e^{-\lambda}) = \Sigma_{-\infty}^\infty a_m[\exp(ime^{-\lambda}) - 1],$$

and hence $K(x) = \Sigma_{-\infty}^\infty a_m E_m(x)$. The result now follows from Lemma 6.30:

$$\int |x|^\alpha |K(x)|\,dx \leq \Sigma_{-\infty}^\infty \|F\|_{C^{(s)}}(1+|m|)^{-s} \int |x|^\alpha |E_m(x)|\,dx$$

$$\leq C_{\alpha,s} \|M\|_{C^{(s)}}(1 + \textstyle\sum_{m\neq 0} |m|^{-s+\alpha+(Q/2)+1}). \quad \#$$

(6.36) LEMMA. Under the same hypotheses as Lemma 6.35, K is C^∞, and for every multiindex I,

$$|X^I K(x)| \leq C_{I,s,\alpha} |x|^{-\alpha} \|M\|_{C^{(s)}},$$

$$|Y^I K(x)| \leq C_{I,s,\alpha} |x|^{-\alpha} \|M\|_{C^{(s)}}.$$

Proof: Let $M_1(\lambda) = e^{\lambda}M(\lambda)$, and let K_1 be the distribution kernel of $M_1(L)$. Then M_1 satisfies the same hypotheses as M, and the $C^{(s)}$ norms of M and M_1 are comparable. Moreover, $M(L) = e^{-L}M_1(L) = M_1(L)e^{-L}$, so $K = h(\cdot,1)*K_1 = K_1*h(\cdot,1)$. Hence K is C^{∞}, and

$$X^I K = K_1 * X^I h(\cdot,1), \qquad Y^I K = Y^I h(\cdot,1)*K_1.$$

Since $h(\cdot,1)\in S$, by Lemmas 1.10 and 6.35 we obtain

$$(1+|x|)^{\alpha}|X^I K(x)| \le \gamma^{\alpha}\int(1+|y|)^{\alpha}\,|K_1(y)|(1+|y^{-1}x|)^{\alpha}|X^I h(y^{-1}x,1)|dy$$

$$\le A_{I,\alpha}\int(1+|y|^{\alpha})|K_1(y)|dy$$

$$\le A_{I,\alpha}(C_{s,0}+C_{s,\alpha})\ \|M_1\|_{C^{(s)}}$$

$$\le C_{I,s,\alpha}\ \|M\|_{C^{(s)}},$$

and similary for $Y^I K(x)$. #

Proof of Theorem 6.25: Suppose M is of class $C^{(s)}$ on $(0,\infty)$ and $\sup_{\lambda>0}|\lambda^j M^{(j)}(\lambda)| \le C < \infty$ for $0 \le j \le s$, and let K be the distribution kernel of $M(L)$. We shall show that if $s > r+(3Q/2)+2$ then $X^I K$ is continuous on $G\backslash\{0\}$ and $|X^I K(x)| \le C_I|x|^{-Q-d(I)}$ for $d(I) \le r$, and likewise for $Y^I K$. The desired result then follows from Theorem 6.10 and Remark 6.12.

Fix $\psi\in C_0^{\infty}([1/2,2])$ with $\psi > 0$ on $(1/2,2)$. For $j\in\mathbb{Z}$ let $\psi_j(x) = \psi(2^{-j}x)$, and then set $\phi_j(x) = \psi_j(x)/\Sigma_i\psi_i(x)$. Thus $\{\phi_j\}_{j\in\mathbb{Z}}$ is a partition of unity on $(0,\infty)$ such that $\operatorname{supp}\phi_j \subset [2^{j-1},2^{j+1}]$ and $\|\phi_j^{(\ell)}\|_\infty \le C2^{-\ell j}$. Let $M_j(\lambda) = M(\lambda)\phi_j(\lambda)$, and let K_j be the distribution kernel of $M_j(L)$. By the hypotheses on M we have

$$\|M_j^{(\ell)}\|_\infty \le C \sum_{i=0}^{\ell} \binom{\ell}{i} 2^{-ij}2^{-(\ell-i)j} = C'2^{-\ell j} \qquad (0 \le \ell \le s),$$

which implies that the functions $\lambda \to M_j(2^j\lambda)$ satisfy the hypotheses of Lemma 6.35 with $C^{(s)}$ norm bounded independently of j. Hence, in view of Lemmas 6.29 and 6.36, if $s > \alpha + (Q/2) + 2$ we have

$$2^{-j(Q+d(I))/2} |X^I K_j(2^{-j/2}x)| \le C_{I,s,\alpha} |x|^{-\alpha},$$

or in other words,

$$(6.37) \qquad |X^I K_j(x)| \le 2^{(j/2)(Q+d(I)-\alpha)} C_{I,s,\alpha} |x|^{-\alpha}.$$

Next, for $x \neq 0$ and $d(I) \le r$ we write

$$X^I K(x) = \Sigma_{-\infty}^{j_0} X^I K_j(x) + \Sigma_{j_0+1}^{\infty} X^I K_j(x),$$

where j_0 is an integer to be determined later. (As we shall see, the series on the right converge uniformly on compact subsets of $G\backslash\{0\}$.) To estimate the terms with $j \le j_0$ we use (6.37) with $\alpha = 0$:

$$(6.38) \qquad \sum_{j\le j_0} |X^I K_j(x)| \le C \sum_{j\le j_0} 2^{(j/2)(Q+d(I))} \le C'2^{(j_0/2)(Q+d(I))}.$$

To estimate the terms with $j > j_0$ we pick $\varepsilon > 0$ such that $\varepsilon < s-r-(3Q/2)-2$ and use (6.37) with $\alpha = Q + d(I) + \varepsilon$:

$$(6.39) \qquad \sum_{j>j_0} |X^I K_j(x)| \le C \sum_{j>j_0} 2^{-\varepsilon j/2} |x|^{-Q-d(I)-\varepsilon} \le C'2^{-\varepsilon j_0/2} |x|^{-Q-d(I)-\varepsilon}$$

Finally, for each $x \neq 0$ we choose j_0 so that $2^{j_0/2} \leq |x|^{-1} \leq 2^{(j_0+1)/2}$. Then the right hand sides of (6.38) and (6.39) are both dominated by $|x|^{-Q-d(I)}$. The same argument also works for $Y^I K$, so the proof is complete. #

Notes and References

The L^p theory, $p > 1$: For the classical theory of singular and fractional integrals on $\mathbb{R}^n$, which has a long history, see Stein [2] and Stein and Weiss [2]. Theorem 6.19 is due to Knapp and Stein [1]. Other conditions which guarantee the L^2 boundedness of singular integral operators on homogeneous groups have been investigated by Goodman [2], Ricci [1], and Strichartz [1]. Once the L^2 theory is established, the boundedness of such operators on L^p $(1 < p < \infty)$ can be proved by showing that they are weak type $(1,1)$: see Fabes and Riviere [1], Coifman and Weiss [1], Korányi and Vági [1], and Strichartz [1]. Our proof of Theorem 6.21 is adapted from an argument of Stein and Zygmund [1].

The H^p theory, $p \leq 1$: For the case of Riesz potentials on $\mathbb{R}^n$, Theorem 6.10 is due to Stein and Weiss [1]; for kernels of type $(0,r)$ on $\mathbb{R}^n$ it is due to Fefferman and Stein [1], but an earlier version is in Stein [2]; for homogeneous kernels on the Heisenberg group it is due to Krantz [3]. See also Calderón and Torchinsky [2], Coifman and Weiss [2], Hemler [1], Mauceri, Picardello, and Ricci [1], and Taibleson and Weiss [1] for related results.

Multipliers: For multipler theorems on $\mathbb{R}^n$, see Stein [2], where earlier references are given; also Calderón and Torchinsky [2], Coifman and Meyer [1] and Taibleson and Weiss [1]. Theorem 6.25 is joint work of Hulanicki and Stein [1]; details of the proof appear here for the first time. deMichele and Mauceri [1] have proved a multiplier theorem for the three-dimensional Heisenberg group which includes Theorem 6.25 for $G = H_1$, $p > 1$. For the case $p > 1$ and $M(\lambda) = \lambda \int_0^\infty e^{-\lambda t} m(t)dt$ where m is bounded on $(0,\infty)$ (which implies that M is analytic and that $\sup_{\lambda>0} |\lambda^j M^{(j)}(\lambda)| < \infty$ for all j), the result is contained in a general theorem of Stein [3], p. 121. Note that by the use of Theorem 3.37 it follows that the multiplier operator $M(L)$ (of Theorem 6.25) is of weak type $(1,1)$, whenever $s > \frac{3}{2}Q + 3$.

CHAPTER 7.

Characterization of H^p by Square Functions: The Lusin and Littlewood-Paley Functions

In this chapter we show that a distribution in H^p can be characterized by analogues to the Lusin and Littlewood-Paley square functions. The material here breaks up into four parts: First, the fact that $f \in H^p$ implies $S_\phi(f) \in L^p$, and its variants, (Theorems 7.7 and 7.8), which follow easily from the results on boundedness of convolution operators obtained in the previous chapter. Second, the converse direction for $1 < p$, (Theorem 7.10), which is proved by a duality argument. Third, the converse direction for $p \leq 1$ which is the most difficult result in this chapter (see Corollary 7.22). Finally there are, as a consequence, corresponding results for stratified groups for square functions fashioned out of the heat semi-group.

Suppose that $f \in \mathcal{S}'$, $0 < \alpha < \infty$, and $\phi \in \mathcal{S}$, and that $\int \phi = 0$. We define the _Lusin function_ (or _area integral_) $S_\phi^\alpha f$ by

$$S_\phi^\alpha f(x) = \left[\int_0^\infty \int_{|x^{-1}y|<\alpha t} |f*\phi_t(y)|^2 t^{-Q-1} dy dt\right]^{1/2}.$$

We shall be mainly concerned with the case $\alpha = 1$, and we set

$$S_\phi f = S_\phi^1 f.$$

Also, suppose that $0 < \lambda < \infty$. We define the _Littlewood-Paley functions_

$g_\phi f$ and $G_\phi^\lambda f$ by

$$g_\phi f(x) = [\int_0^\infty |f*\phi_t(y)|^2 t^{-1} dt]^{1/2},$$

$$G_\phi^\lambda f(x) = [\int_0^\infty \int_G |f*\phi_t(y)|^2 (\frac{t}{t+|x^{-1}y|})^{2\lambda} t^{-Q-1} dy dt]^{1/2}.$$

The functions $g_\phi f$, $S_\phi f$, and $G_\phi^\lambda f$ bear roughly the same relations to each other as the radial, non-tangential, and tangential maximal functions $M_\phi^0 f$, $M_\phi f$, and $T_\phi^\lambda f$, and they may be considered as L^2 analogues of these maximal functions. The principal object of this chapter is to show how H^p can be characterized in terms of Lusin and Littlewood-Paley functions. We begin, however, by deriving some inequalities relating the latter functions.

(7.1) THEOREM. <u>Suppose</u> $0 < p \leq 2$ <u>and</u> $0 < \beta < \alpha < \infty$. If $S_\phi^\beta f \in L^p$ <u>then</u> $S_\phi^\alpha f \in L^p$, <u>and there is a constant</u> C, <u>depending only on</u> p, <u>such that</u>

$$\|S_\phi^\alpha f\|_p \leq C(\alpha/\beta)^{Q/p} \|S_\phi^\beta f\|_p.$$

<u>Proof</u>: For simplicity we shall assume that $\beta = 1$; the argument is the same for general β. If $p = 2$, we have

$$\|S_\phi^\alpha f\|_2^2 = \int_G \int_0^\infty \int_{|x^{-1}y|<\alpha t} |f*\phi_t(y)|^2 t^{-Q-1} dy dt dx$$

$$= \int_0^\infty \int_G |f*\phi_t(y)|^2 \, |B(\alpha t,y)| t^{-Q-1} dy dt$$

$$= \alpha^Q \int_0^\infty \int_G |f*\phi_t(y)|^2 \, |B(t,y)| t^{-Q-1} dy dt$$

$$= \alpha^Q \|S_\phi f\|_2^2.$$

Henceforth we assume that $p < 2$. We shall need the following modification of the Hardy-Littlewood maximal function:

$$\tilde{M}f(x) = \sup_{|x^{-1}y|<\alpha t} t^{-Q}\left|\int_{B(t,y)} f(z)dz\right|$$

$$= M_\psi f(x), \quad \text{where} \quad \psi = \alpha^Q \chi_{B(1/\alpha,0)}.$$

By the maximal theorem (2.4), for $\alpha \geq 1$ we have

$$(7.2) \qquad |\{x : \tilde{M}f(x) > \lambda\}| \leq C\alpha^Q \|f\|_1/\lambda,$$

with C independent of f, α, and λ. Now for $\lambda > 0$, let

$$E_\lambda = \{x : S_\phi f(x) > \lambda\}, \qquad \tilde{E}_\lambda = \{x : \tilde{M}(\chi_{E_\lambda}) > 1/2\}.$$

We claim that

$$(7.3) \qquad \int_{\tilde{E}_\lambda^c} |S_\phi^\alpha f|^2 \leq 2\alpha^Q \int_{E_\lambda^c} |S_\phi f|^2.$$

Assuming this for the moment, we complete the proof of the theorem. By Chebyshev's inequality, (7.2), and (7.3),

$$|\{x : S_\phi^\alpha f(x) > \lambda\}| \leq |\tilde{E}_\lambda| + |\tilde{E}_\lambda^c \cap \{x : S_\phi^\alpha f(x) > \lambda\}|$$

$$\leq |\tilde{E}_\lambda| + \lambda^{-2} \int_{\tilde{E}_\lambda^c} |S_\phi^\alpha f|^2$$

$$\leq C\alpha^Q |E_\lambda| + 2\alpha^Q \lambda^{-2} \int_{E_\lambda^c} |S_\phi f|^2$$

$$\leq C\alpha^Q [\,|E_\lambda| + \lambda^{-2} \int_0^\lambda t|E_t|dt]$$

Hence,

$$\int |S_\phi^\alpha f|^p = \int_0^\infty p\lambda^{p-1}|\{x : S_\phi^\alpha f(x) > \lambda\}|d\lambda$$

$$\leq C\alpha^Q\Big[\int_0^\infty p\lambda^{p-1}|E_\lambda|d\lambda + \int_0^\infty \int_0^\lambda \lambda^{p-3}t|E_t|dtd\lambda\Big]$$

$$\leq C\alpha^Q\Big[\int_0^\infty |S_\phi f|^p + \int_0^\infty \int_t^\infty \lambda^{p-3}t|E_t|d\lambda dt\Big]$$

$$\leq C\alpha^Q\Big[\int_0^\infty |S_\phi f|^p + (p-2)^{-1}\int_0^\infty t^{p-1}|E_t|dt\Big]$$

$$= C'\alpha^Q \int |S_\phi f|^p.$$

It therefore remains to prove (7.3). Setting $\rho(y) = \inf\{|y^{-1}z| : z \in \tilde{E}_\lambda^c\}$, we have

$$\int_{\tilde{E}_\lambda^c} |S_\phi^\alpha f|^2 = \int_{\tilde{E}_\lambda^c} \int_0^\infty \int_{|x^{-1}y|<\alpha t} |f*\phi_t(y)|^2 t^{-Q-1}dydtdx$$

$$= \int_0^\infty \int_{\rho(y)<\alpha t} |f*\phi_t(y)|^2|\tilde{E}_\lambda^c \cap B(\alpha t,y)|t^{-Q-1}dydt$$

$$\leq \int_0^\infty \int_{\rho(y)<\alpha t} |f*\phi_t(y)|^2(\alpha t)^Q t^{-Q-1}dydt.$$

Observe that if $x \in \tilde{E}_\lambda^c \cap B(\alpha t,y)$ then

$$t^{-Q}|E_\lambda \cap B(t,y)| = t^{-Q}\int_{B(t,y)} \chi_{E_\lambda}(z)dz \leq \tilde{M}(\chi_{E_\lambda})(x) \leq 1/2,$$

and hence

$$t^{-Q}|E_\lambda^c \cap B(t,y)| = t^{-Q}[t^Q - |E_\lambda \cap B(t,y)|] \geq 1/2.$$

Now if $\rho(y) < \alpha t$ then $\tilde{E}_\lambda^c \cap B(\alpha t, y)$ is nonempty, so

$$\int_{\tilde{E}_\lambda^c} |S_\phi^\alpha f|^2 \leq \int_0^\infty \int_{\rho(y)<\alpha t} |f*\phi_t(y)|^2 \alpha^Q 2 |E_\lambda^c \cap B(t,y)| t^{-Q-1} dy dt$$

$$\leq 2\alpha^Q \int_0^\infty \int_G |f*\phi_t(y)|^2 |E_\lambda^c \cap B(t,y)| t^{-Q-1} dy dt$$

$$= 2\alpha^Q \int_{E_\lambda^c} |S_\phi f|^2. \quad \#$$

(7.4) COROLLARY. _Suppose_ $0 < p \leq 2$ _and_ $\lambda > Q/p$. _If_ $S_\phi f \in L^p$ _then_ $G_\phi^\lambda f \in L^p$, _and_ $\|G_\phi^\lambda f\|_p \leq C_{\lambda,p} \|S_\phi f\|_p$.

Proof: In the integral defining $G_\phi^\lambda f(x)$, break up the integration over G into integration over the sets where $|x^{-1}y| < t$ and where $2^k t \leq |x^{-1}y| < 2^{k+1} t$ $(k = 0,1,2,\ldots)$. It is then clear that

$$(G_\phi^\lambda f(x))^2 \leq (S_\phi f(x))^2 + \Sigma_0^\infty (1+2^k)^{-2\lambda} (S_\phi^{2^{k+1}} f(x))^2.$$

Raise both sides to the power $p/2$ and apply Theorem 7.1:

$$\int |G_\phi^\lambda f|^p \leq \int |S_\phi f|^p [1 + C\Sigma_0^\infty (1+2^k)^{-\lambda p} 2^{kQ}].$$

The last series converges when $\lambda > Q/p$, so we are done. #

(7.5) THEOREM. _If_ $\psi \in \mathcal{S}$ _and_ $\lambda > 0$, _there exists_ $C > 0$ _such that for all_ $f \in \mathcal{S}'$, $x \in G$, _and_ $\phi \in \mathcal{S}$ _with_ $\int \phi = 0$,

$$S_{\phi*\psi} f(x) \leq C G_\phi^\lambda f(x).$$

Proof: We have

$$(S_{\phi*\psi} f(x))^2 = \int_0^\infty \int_{|x^{-1}y|<t} |f*\phi_t*\psi_t(y)|^2 t^{-Q-1} dydt$$

$$= \int_0^\infty \int_{|x^{-1}y|<t} \left|\int_G f*\phi_t(z)\psi_t(z^{-1}y)dz\right|^2 t^{-Q-1} dydt.$$

In the inner integral, multiply and divide by $(1+t^{-1}|y^{-1}z|)^\lambda$ and use the Schwarz inequality:

$$(S_{\phi*\psi} f(x))^2 \leq \int_0^\infty \int_{|x^{-1}y|<t} \left[\int_G |f*\phi_t(z)|^2 \left(\frac{t}{t+|y^{-1}z|}\right)^{2\lambda} dz\right] \times$$

$$\left[\int_G \left(\frac{t+|y^{-1}z|}{t}\right)^{2\lambda} \left|\psi\left(\frac{z^{-1}y}{t}\right)\right|^2 t^{-2Q} dz\right] t^{-Q-1} dydt.$$

Let $C_1 = \int_G (1+|w|)^{2\lambda} |\psi(w)|^2 dw$: then

$$(S_{\phi*\psi} f(x))^2 \leq C_1 \int_0^\infty \int_{|x^{-1}y|<t} \int_G |f*\phi_t(z)|^2 \left(\frac{t}{t+|y^{-1}z|}\right)^{2\lambda} dz t^{-2Q-1} dydt.$$

However, if $|x^{-1}y| < t$ then $|x^{-1}z| < \gamma(t+|y^{-1}z|)$, and hence

$$t+|x^{-1}z| \leq (\gamma+1)(t+|y^{-1}z|).$$

Therefore,

$$(S_{\phi*\psi} f(x))^2 \leq C_2 \int_0^\infty \int_G \int_{|x^{-1}y|<t} |f*\phi_t(z)|^2 \left(\frac{t}{t+|x^{-1}z|}\right)^{2\lambda} t^{-2Q-1} dydzdt$$

$$= C_2 (G_\lambda^\phi f(x))^2. \quad \#$$

(7.6) COROLLARY. *If* $S_\phi f \in L^p$ $(0 < p \leq 2)$ *then* $S_{\phi*\psi} f \in L^p$ *and* $\|S_{\phi*\psi} f\|_p \leq C_{p,\psi} \|S_\phi f\|_p$.

Proof: Combine Theorem 7.5 and Corollary 7.4. #

We are now ready to explore the relationship between H^p spaces and the Lusin and Littlewood-Paley functions.

(7.7) THEOREM. *If* $\phi \in \mathcal{S}$ *and* $\int \phi = 0$, *the map* $f \to g_\phi f$ *is bounded from* H^p *to* L^p *for* $0 < p < \infty$.

Proof: First, suppose $p = 2$. Since the map $f \to g_\phi f$ is subadditive, it suffices to prove that $\|g_\phi f\|_2 \leq C\|f\|_2$ for $f \in \mathcal{S}$. Let $\tilde{\phi}(x) = \overline{\phi(x^{-1})}$: then

$$\|g_\phi f\|_2^2 = \lim_{\varepsilon \to 0, A \to \infty} \int_\varepsilon^A \int_G |f*\phi_t(x)|^2 t^{-1} dxdt$$

$$= \lim \int_\varepsilon^A \int_G \int_G \int_G f(y)\phi_t(y^{-1}x)\overline{\phi_t(u^{-1}x)f(u)}t^{-1} dudydxdt$$

$$= \lim \int_\varepsilon^A \int_G \int_G f(y)\phi_t*\tilde{\phi}_t(y^{-1}u)\overline{f(u)}t^{-1} dydudt$$

$$= \lim \int_G f*K_\varepsilon^A(u)\overline{f(u)}du$$

where

$$K_\varepsilon^A(y) = \int_\varepsilon^A (\phi*\tilde{\phi})_t(y)t^{-1}dt.$$

By Theorem 1.65, as $\varepsilon \to 0$ and $A \to \infty$, K_ε^A converges in $\mathcal{S}'$ to a distribution K which is C^∞ on $G\setminus\{0\}$ and homogeneous of degree $-Q$. Also, since $f \in S$, $f*K_\varepsilon^A \to f*K$ in S'. Hence, by Theorem 6.19,

$$\|g_\phi f\|_2^2 = \int f*K(u)\overline{f(u)}du \leq \|f*K\|_2\|f\|_2 \leq C\|f\|_2^2.$$

Next, let X denote the Hilbert space $L^2((0,\infty),dt/t)$. Define the X-valued distribution Φ on G by

$$\langle f,\Phi\rangle(t) = \int f(x)\phi_t(x)dx \qquad (f \in S).$$

$\langle f,\Phi\rangle$ does belong to X because it is $O(t^{-Q})$ as $t \to \infty$ and (by Proposition 1.49, since $\int\phi = 0$) $O(t)$ as $t \to 0$. Moreover, $g_\phi f(x) = \|f*\Phi(x)\|_X$, so we have just proved that the map $f \to f*\Phi$ is bounded from L^2 to L_X^2. Furthermore, Φ is given on $G\setminus\{0\}$ by integration against the smooth X-valued function $\Phi(x)(t) = \phi_t(x)$ which satisfies

$$\begin{aligned}\|Y^I\Phi(x)\|_X^2 &= \int_0^\infty t^{-2Q-2d(I)}|Y^I\phi(x/t)|^2 dt/t \\ &\leq C\int_0^{|x|} t^{-2Q-2d(I)-1}|x/t|^{-2N}dt + C\int_{|x|}^\infty t^{-2Q-2d(I)-1}dt \\ &\leq C'|x|^{-2Q-2d(I)}\end{aligned}$$

(where we have chosen $N > Q+d(I)$). Therefore, Φ is an X-valued (i.e., $B(\mathbb{C},X)$-valued) kernel of type $(0,r)$ for all r, so by Theorem 6.20 we conclude that the map $f \to f*\Phi$ is bounded from H^p to H_X^p for $0 < p < \infty$. In particular, for $p > 1$ we have

$$\|g_\phi f\|_p^p = \int \|f*\Phi(x)\|_X^p dx \leq C_p\|f\|_p^p.$$

The case $p \leq 1$ requires a slight extra twist. If $f \in S \cap H^p$ and $\psi \in S$ then

$$\int \|(f*\Phi)*\psi_\varepsilon(x)\|_X^p dx \leq \int M_\psi(f*\Phi)(x)^p dx \leq C\rho^p(f).$$

But if we choose ψ to satisfy $\int \psi = 1$, we also have $(f*\Phi)*\psi_\varepsilon \to f*\Phi$ pointwise as $\varepsilon \to 0$, so by Fatou's lemma,

$$\|g_\phi f\|_p^p = \int \|f*\Phi(x)\|_X^p dx \leq C\rho^p(f).$$

Since $S \cap H^p$ is dense in H^p by Theorem 3.33, a simple limiting argument completes the proof. #

(7.8) THEOREM. <u>If</u> $\phi \in S$ <u>and</u> $\int \phi = 0$, <u>and</u> $0 < \alpha < \infty$, <u>the map</u> $f \to S_\phi^\alpha f$ <u>is bounded from</u> H^p <u>to</u> L^p <u>for</u> $0 < p < \infty$.

<u>Proof</u>: For simplicity we assume that $\alpha = 1$. For $p = 2$, we have

$$\|S_\phi f\|_2^2 = \int_G \int_0^\infty \int_{|x^{-1}y|<t} |f*\phi_t(y)|^2 t^{-Q-1} dy dt dx$$

$$= \int_0^\infty \int_G |f*\phi_t(y)|^2 |B(t,y)| t^{-Q-1} dy dt$$

$$= \int_0^\infty \int_G |f*\phi_t(y)|^2 t^{-1} dt$$

$$= \|g_\phi f\|_2^2,$$

so the result follows from Theorem 7.7.

The argument for $p \neq 2$ now follows the same lines as the proof of Theorem 7.7. To wit, we set

$$X = L^2(\{(y,t) : 0 < |y| < t < \infty\}, t^{-Q-1}dydt)$$

and we define the X-valued distribution Φ on G by

$$<f,\Phi>(y,t) = \int f(x)\phi_t(xy)dx.$$

Then $S_\phi f(x) = \|f*\Phi(x)\|_X$, and Φ is an X-valued kernel of type $(0,r)$ for all r, since

$$\|Y^I\Phi(x)\|_X^2 = \int_0^\infty \int_{|x^{-1}y|<t} t^{-2d(I)-3Q-1}|Y^I\phi(y/t)|^2dydt$$

$$\leq C\int_{t<|x|/2\gamma}\int_{|x^{-1}y|<t} t^{-2d(I)-3Q-1}|y/t|^{2N}dydt$$

$$+ C\int_{t\geq|x|/2\gamma}\int_{|x^{-1}y|<t} t^{-2d(I)-3Q-1}dydt.$$

The second term on the right is a constant times $|x|^{-2Q-2d(I)}$, and for the first we observe that if $|x^{-1}y| < |x|/2\gamma$ then $|y| > |x|/2$, whence it follows (if we take $N > Q+d(I)$) that the first term is also dominated by $|x|^{-2Q-2d(I)}$. The desired result now follows as before; details are left to the reader. #

We have now shown that if $f \in H^p$ then $g_\phi f \in L^p$ and $S_\phi f \in L^p$ for any $\phi \in S$ with $\int \phi = 0$. It is obvious that the converse cannot hold without some additional restrictions on ϕ and f. For example if $G = \mathbb{R}$

and the Fourier transform of ϕ is supported in $(0,\infty)$, then $f*\phi_t$ vanishes identically (and hence so do $g_\phi f$ and $S_\phi f$) for any $f \in \mathcal{S}'$ whose Fourier transform is supported in $(-\infty,0)$. Moreover, we must in any event require that $\int \phi = 0$ for the integrals defining $g_\phi f$ and $S_\phi f$ to have much hope of converging, and this implies that $g_\phi f = S_\phi f = 0$ whenever f is a constant function.

We shall make use of the terminology and propositions in Chapter 1, Section E. We call attention, in particular, to Theorems 1.61 and 1.62, which guarantee the existence of Schwartz class functions which satisfy the hypotheses of the following theorems. We shall restrict our attention to distributions which vanish weakly at infinity, in order to exploit Theorem 1.64. The following elementary result provides reassurance that this restriction is reasonable:

(7.9) PROPOSITION. *If* $f \in H^p$ $(0 < p < \infty)$, *then* f *vanishes weakly at infinity. More precisely, for any* $\phi \in \mathcal{S}$,

$$\|f*\phi_t\|_\infty \leq \|M_\phi f\|_p \, t^{-Q/p}.$$

Proof: We observe that $|f*\phi_t(x)| \leq M_\phi f(y)$ whenever $|x^{-1}y| < t$. Hence

$$|f*\phi_t(x)|^p \leq |B(t,x)|^{-1} \int_{B(t,x)} M_\phi f(y)^p dy \leq t^{-Q}\|M_\phi f\|_p^p. \quad \#$$

(7.10) THEOREM. *Suppose* $1 < p < \infty$, $\phi^1,\ldots,\phi^N,\psi^1,\ldots,\psi^N \in \mathcal{S}$, $\int \phi^j = \int \psi^j = 0$ *for* $1 \leq j \leq N$, *and* $\Sigma_1^N \int_0^\infty \phi_t^j * \psi_t^j dt/t = \delta$ *in the sense of* (1.59'). *If* $f \in \mathcal{S}'$ *vanishes weakly at infinity and either* $g_{\phi^j} f \in L^p$ *for*

$1 \leq j \leq N$ <u>or</u> $S_{\phi^j} f \in L^p$ <u>for</u> $1 \leq j \leq N$, <u>then</u> $f \in L^p$ <u>and</u>

$$\|f\|_p \leq C_p \Sigma_1^N \|g_{\phi^j} f\|_p, \qquad \|f\|_p \leq C_p \Sigma_1^N \|S_{\phi^j} f\|_p.$$

<u>Proof</u>: If $0 < \varepsilon < A < \infty$, let

$$f_\varepsilon^A = \Sigma_1^N \int_\varepsilon^A f*\phi_t^j * \psi_t^j \, dt/t.$$

Then by Theorem 1.64, $f_\varepsilon^A \to f$ in S' as $\varepsilon \to 0$, $A \to \infty$. If we set $\tilde{\psi}^j(x) = \overline{\psi^j(x^{-1})}$, for any $\eta \in S$ we have

$$\left|\int f(x)\overline{\eta(x)}dx\right| = \left|\lim_{\varepsilon \to 0, A \to \infty} \int f_\varepsilon^A(x)\overline{\eta(x)}dx\right|$$

$$= \left|\lim \Sigma_1^N \int_G \int_\varepsilon^A \int_G f*\phi_t(u)\psi_t(u^{-1}x)\overline{\eta(x)}t^{-1} du\,dt\,dx\right|$$

$$= \left|\lim \Sigma_1^N \int_G \int_\varepsilon^A f*\phi_t^j(u)\overline{\eta*\tilde{\psi}_t^j(u)}t^{-1} dt\,du\right|$$

$$\leq \lim \Sigma_1^N \int_G \left[\int_\varepsilon^A |f*\phi_t^j(u)|^2 t^{-1} dt\right]^{1/2} \left[\int_\varepsilon^A |\eta*\tilde{\psi}_t^j(u)|^2 t^{-1} dt\right]^{1/2} du$$

$$= \Sigma_1^N \int_G (g_{\phi^j} f(u))(g_{\tilde{\psi}^j} \eta(u))du.$$

Hence if $g_{\phi^j} f \in L^p$ and q is the conjugate exponent to p, by Theorem 7.7 we have

$$\|f\|_p = \sup_{\eta \in S} \left|\int f\bar{\eta}\right| / \|\eta\|_q \leq \Sigma_1^N \|g_{\phi^j} f\|_p \|g_{\tilde{\psi}^j} \eta\|_q / \|\eta\|_q$$

$$\leq C\Sigma_1^N \|g_{\phi^j} f\|_p.$$

A similar calculation shows that

$$\left|\int f(x)\overline{\eta(x)}dx\right| \leq \Sigma_1^N \int (S_{\phi^j}f(u))(S_{\tilde{\psi}^j}\eta(u))du$$

and hence that $S_{\phi^j}f \in L^p$ implies $f \in L^p$. #

To obtain a similar result for $p \leq 1$ one must impose additional conditions on the functions ψ^j. One version of this result is the following theorem; another one will be given in Corollary 7.22. Both of these deal only with the Lusin functions; however, see also Theorem 7.28.

(7.11) THEOREM. Suppose $0 < p \leq 1$, $\phi^1,\ldots,\phi^N,\psi^1,\ldots,\psi^N \in \mathcal{S}$, $\int\phi^j = 0$ for $1 \leq j \leq N$, and $\Sigma_1^N \int_0^\infty \phi_t^j * \psi_t^j dt/t = \delta$ in the sense of (1.59'). Suppose moreover that for $1 \leq j \leq N$, $\text{supp}\ \psi^j \subset B(1,0)$ and $\int \psi^j P = 0$ for all $P \in P_a$, where a is p-admissible and $a \geq Q/2$. If $f \in \mathcal{S}'$ vanishes weakly at infinity and $S_{\phi^j}f \in L^p$ for $1 \leq j \leq N$, then $f \in H^p$, and $\rho^p(f) \leq C_p \Sigma_1^N \|S_{\phi^j}f\|_p^p$.

The proof of this theorem will require considerable preparation. To begin with, by Theorem 1.64 we have

$$(7.12) \qquad f = \Sigma_1^N \lim_{\varepsilon\to 0, A\to\infty} \int_\varepsilon^A f*\phi_t^j*\psi_t^j dt/t = \Sigma_1^N F^j.$$

Our strategy will be to fashion an atomic decomposition for f out of this representation of f. More precisely, we shall construct an atomic decomposition for each term F^j on the right of (7.12). We fix j, and henceforth we omit writing the superscript j's: thus,

$$(7.13) \qquad F = \lim_{\varepsilon\to 0, A\to\infty} \int_\varepsilon^A f*\phi_t*\psi_t dt/t.$$

The first step is the construction of an analogue on G of the familiar decomposition of $\mathbb{R}^n$ into dyadic cubes. We choose, once and for all, a maximal family $\{B(1/2\gamma, x_j)\}_{j=1}^{\infty}$ of disjoint ball of radius $1/2\gamma$, and for $k \in \mathbb{Z}$ we set

$$B_j^k = B((2\gamma)^k, (2\gamma)^k x_j).$$

(The reason for using $(2\gamma)^k$ rather than 2^k will appear in the proof of Lemma 7.19 below.) Then for each k, the balls $\{B_j^k\}_{j=1}^{\infty}$ cover G. Indeed, if $x \in G$, by maximality there exists j such that the distance from $(2\gamma)^{-k}x$ to $B(1/2\gamma, x_j)$ is less than $1/2\gamma$. But then

$$|((2\gamma)^k x_j)^{-1} x| < (2\gamma)^k \gamma((2\gamma)^{-1} + (2\gamma)^{-1}) = (2\gamma)^k,$$

so that $x \in B_j^k$.

(7.14) LEMMA. *Given* $C \geq 1$, *let* $\tilde{B}_j^k = B(C(2\gamma)^k, (2\gamma)^k x_j)$. *Then there exist integers* N, N', N'' *depending only on* C *such that for each* $k \in \mathbb{Z}$,

(a) *no point of* G *belongs to more than* N *of the balls* $\tilde{B}_j^k$ $(j = 1,2,3,\ldots)$;

(b) *if* $i = 1,2,3,\ldots$ *and* $k' \geq k$, *there are at most* N' *values of* j *for which* $\tilde{B}_i^k \cap \tilde{B}_j^{k'} \neq \emptyset$;

(c) *if* $i = 1,2,3,\ldots$ *and* $k' < k$, *there are at most* $N''(2\gamma)^{(k-k')Q}$ *values of* j *for which* $\tilde{B}_i^k \cap \tilde{B}_j^{k'} \neq \emptyset$.

Proof: Since B_j^k is obtained from B_j^0 by dilation by the factor $(2\gamma)^k$, it suffices to assume that $k = 0$. First, if $x \in \bigcap_{\ell=1}^N \tilde{B}_{j_\ell}^0$ then $\bigcup_1^N \tilde{B}_{j_\ell}^0 \subset B(4\gamma^2 C, x)$. But then since $B(1/2\gamma, x_j) \subset B_j^0$ we have

$$N(2\gamma)^{-Q} = |\bigcup_1^N B(1/2\gamma, x_{j_\ell})| \leq |B(4\gamma^2 C, x)| = (4\gamma^2 C)^Q,$$

and hence $N \leq (8\gamma^3 C)^Q$, which proves (a). Next, if $k' \geq 0$ and $\tilde{B}_i^0 \cap \tilde{B}_j^{k'} \neq \emptyset$, then

$$x_i \in B(\gamma C((2\gamma)^{k'} + 1), (2\gamma)^{k'} x_j) \subset B(2\gamma C(2\gamma)^{k'}, (2\gamma)^{k'} x_j).$$

By part (a) with C replaced by $2\gamma C$, this can happen for at most $N' = (16\gamma^4 C)^Q$ values of j, which proves (b). Likewise, if $k' < 0$ and $\tilde{B}_i^0 \cap \tilde{B}_{j_\ell}^{k'} \neq \emptyset$ for $\ell = 1, \ldots, M$ then $(2\gamma)^{k'} x_{j_\ell} \in B(2\gamma C, x_i)$, hence

$$x_{j_\ell} \in B((2\gamma)^{1-k'} C, (2\gamma)^{-k'} x_i),$$

so that

$$B(1/2\gamma, x_{j_\ell}) \subset B(2^{-1} + (2\gamma)^{1-k'} \gamma C,\ (2\gamma)^{-k'} x_i) \subset B((2\gamma)^{2-k'} C, (2\gamma)^{-k'} x_i).$$

But then

$$M(2\gamma)^{-Q} = |\bigcup_1^M B(1/2\gamma, x_{j_\ell})| \leq |B((2\gamma)^{2-k'} C, (2\gamma)^{-k'} x_i)| = (2\gamma)^{(2-k')Q} C^Q,$$

so that $M \leq (4\gamma^2 C)^Q (2\gamma)^{-k'Q}$. This proves (c). #

Next, let

$$\zeta_j^k = \chi_{B_j^k} / \sum_{\ell=1}^\infty \chi_{B_\ell^k}.$$

The sum in the denominator always assumes one of the values $1,2,\ldots,N$ where N is as in Lemma 7.14, and we have $\Sigma_j \zeta_j^k(x) = 1$ for all $x \in G$. Also, let

$$I^k = \{t \in \mathbb{R} : (2\gamma)^{k+1} < t \leq (2\gamma)^{k+2}\}.$$

Then $\Sigma_{j,k}\zeta_j^k(x)\chi_{I^k}(t) = 1$ for all $(x,t) \in G \times (0,\infty)$.

At this point we change our notation slightly. let

$$\mathcal{B} = \{B_j^k : 1 \leq j < \infty, -\infty < k < \infty\}.$$

Henceforth we denote elements of $\mathcal{B}$ simply by B, dropping the indices j and k (which will be used with other meanings later on). Also, if $B = B_j^k = B(r,x) \in \mathcal{B}$ we shall write

$$\zeta_B = \zeta_j^k, \quad I_B = I^k, \quad B^* = B(5\gamma^3 r,x).$$

Now, returning to our distributions f and F, for each $B \in \mathcal{B}$ we set

$$(7.15) \qquad F_B(x) = \int_{I_B}\int_B \zeta_B(y) f*\phi_t(y)\psi_t(y^{-1}x)dydt/t.$$

From (7.13) and the properties of ζ_B and I_B, it follows that

$$(7.16) \qquad F = \Sigma_{B\in\mathcal{B}} F_B,$$

the sum converging in $\mathcal{S}'$. We make two elementary observations concerning F_B:

(i) F_B is a C^∞ function supported in B^*: this follows since ψ is C^∞ and the integrand in (7.15) vanishes unless $B \cap B(t,x) \neq \emptyset$ for some $t \in I_B$ (because $\operatorname{supp}\psi \subset B(1,0)$).

(ii) $\int F_B P = 0$ for all $P \in \mathcal{P}_a$, because the same is true of ψ.

Hence the F_B's are constant multiples of (p,∞,a)-atoms. However, the equation (7.16) is not an atomic decomposition of f, because the norms of the F_B's do not add up properly. We must be more subtle.

For $B \in \mathcal{B}$, let us define

$$S_B = [\int_{I_B} \int_B |f*\phi_t(y)|^2 dydt/t]^{1/2}.$$

(7.17) LEMMA. For all $B \in \mathcal{B}$ and all multiindices J,

$$\|X^J F_B\|_\infty \leq CS_B|B|^{-(1/2)-(d(J)/Q)},$$

where C depends on J but not on B.

Proof: By the Schwarz inequality,

$$|X^J F_B(x)| = |\int_{I_B} \int_B \zeta_B(y) f*\phi_t(y) X^J \psi(y^{-1}x) dydt/t|$$

$$\leq S_B[\int_{I_B} \int_B |X^J \psi_t(y^{-1}x)|^2 dydt/t]^{1/2}.$$

But $X^J \psi_t(z) = t^{-Q-d(J)}(X^J\psi)(z/t)$. Hence if we set $\alpha = 2\gamma|B|^{1/Q}$, the expression in square brackets is bounded by

$$\int_\alpha^{2\gamma\alpha} \int_G |X^J \psi((y^{-1}x)/t)|^2 \; dy t^{-2Q-2d(J)-1} dt$$

$$= \int_G |X^J\psi(z)|^2 dz \int_\alpha^{2\gamma\alpha} t^{-Q-2d(J)-1} dt$$

$$= C\alpha^{-Q-2d(J)} = C'|B|^{-1-(2d(J)/Q)}. \quad \#$$

Now we are ready to bring in the Lusin function $S_\phi f$. For $k \in \mathbb{Z}$, let

$$\Omega_k = \{x : S_\phi f(x) > (2\gamma)^k\},$$

and

$$\mathcal{B}_k = \{B \in \mathcal{B} : |B \cap \Omega_k| > |B|/2,\ |B \cap \Omega_{k+1}| \leq |B|/2\}.$$

Since the Ω_k's are a nested family of sets whose union is G and whose intersection has measure zero, each $B \in \mathcal{B}$ belongs to precisely one $\mathcal{B}_k$.

(7.18) LEMMA. <u>There exists</u> $C > 0$, <u>independent of</u> k, <u>such that</u>

$$\Sigma_{B\in\mathcal{B}_k} S_B^2 \leq C(2\gamma)^{2k}|\Omega_k|.$$

<u>Proof</u>: Let $\tilde{\Omega}_k = \{x : M_{HL}(\chi_{\Omega_k})(x) > 1/2\}$ (M_{HL} = Hardy-Littlewood maximal function). Then $\Omega_k \subset \tilde{\Omega}_k$, and by the maximal theorem (2.4), $|\tilde{\Omega}_k| \leq C|\Omega_k|$. Now

$$\int_{\tilde{\Omega}_k\setminus\Omega_{k+1}} (S_\phi f)^2 \leq (2\gamma)^{2(k+1)}|\tilde{\Omega}_k| \leq C(2\gamma)^{2k}|\Omega_k|,$$

so it suffices to show that

$$\Sigma_{B\in\mathcal{B}_k} S_B^2 \leq C\int_{\Omega_k\setminus\Omega_{k+1}} (S_\phi f)^2.$$

However,

$$\int_{\tilde{\Omega}_k\setminus\Omega_{k+1}} (S_\phi f(x))^2 dx = \int_0^\infty \int_G |f*\phi_t(y)|^2\ |\{x \in \tilde{\Omega}_k\setminus\Omega_{k+1} : |y^{-1}x| < t\}|t^{-Q-1}dydt.$$

We claim that there exists $C > 0$ such that if $B \in \mathcal{B}_k$, $y \in B$, and $t \in I_B$,

$$|\{x \in \tilde{\Omega}_k \backslash \Omega_{k+1} : |y^{-1}x| < t\}| \geq Ct^Q.$$

Once this is established, we are done. Indeed, since $B \times I_B$ and $B' \times I_{B'}$ are disjoint unless B and B' have the same size, by Lemma 7.14 we have

$$\int_{\tilde{\Omega}_k \backslash \Omega_{k+1}} (S_\phi f(x))^2 dx \geq N^{-1} \Sigma_{B \in \mathcal{B}_k} \int_{I_B} \int_B |f * \phi_t(y)|^2 \times$$

$$|\{x \in \tilde{\Omega}_k \backslash \Omega_{k+1} : |y^{-1}x| < t\}| t^{-Q-1} dy dt$$

$$\geq C' \Sigma_{B \in \mathcal{B}_k} \int_{I_B} \int_B |f * \phi_t(y)|^2 dy dt/t$$

$$= C' \Sigma_{B \in \mathcal{B}_k} S_B^2.$$

It thus remains to prove the claim. If $B \in \mathcal{B}_k$ then $|B \cap \Omega_k| > |B|/2$, hence $M_{HL}(\chi_{\Omega_k})(x) > 1/2$ for $x \in B$, that is $B \subset \tilde{\Omega}_k$. Also, $|B \cap \Omega_{k+1}| \leq |B|/2$, so $|B \backslash \Omega_{k+1}| \geq |B|/2$. Thus if $y \in B$ and $t \in I_B$,

$$\{x \in \tilde{\Omega}_k / \Omega_{k+1} : |y^{-1}x| < t\} \supset \{x \in \tilde{\Omega}_k \backslash \Omega_{k+1} : |y^{-1}x| < 2\gamma |B|^{1/Q}\}$$

$$\supset B \cap (\tilde{\Omega}_k \backslash \Omega_{k+1}) = B \backslash \Omega_{k+1},$$

and hence

$$|\{x \in \tilde{\Omega}_k \backslash \Omega_{k+1} : |y^{-1}x| > t\}| \geq |B \backslash \Omega_{k+1}| \geq |B|/2 \geq (t/4\gamma^2)^Q/2. \quad \#$$

Next, we define a partial ordering on $\mathcal{B}_k$ as follows: we say that $B \preccurlyeq B'$ if either (a) $B = B'$ or (b) there exist $B_0, \ldots, B_m \in \mathcal{B}_k$ with

$B_0 = B$, $B_m = B'$, such that $|B_{j-1}| < |B_j|$ and $B_{j-1} \cap B_j \neq \emptyset$ for $j = 1,\ldots,m$. (If $G = \mathbb{R}^n$ and we replace $\mathcal{B}$ by the family of dyadic cubes, this partial ordering is simply inclusion.) Let $\{B^\ell\}$ be an enumeration of the maximal elements of $\mathcal{B}_k$ with respect to this ordering. Since $B \in \mathcal{B}_k$ implies $|B| < 2|\Omega_k|$ and $B \prec B'$ implies $B = B'$ or $|B'| \geq (2\gamma)^Q|B|$, there are no infinite increasing sequences $B_1 \prec B_2 \prec B_3 \cdots$ with the B_i's all distinct. It follows that each $B \in \mathcal{B}_k$ satisfies $B \prec B^\ell$ for some ℓ. For each $B \in \mathcal{B}_k$ we choose such an ℓ once and for all, and we denote by $\mathcal{B}_k^\ell$ the set of all $B \in \mathcal{B}_k$ for which B^ℓ is the chosen maximal element. Thus $\mathcal{B}_k$ is the disjoint union of the $\mathcal{B}_k^\ell$'s.

(7.19) LEMMA. _If_ $B \in \mathcal{B}_k^\ell$ _then_ $B \subset (B^\ell)^*$.

Proof: If $B \in \mathcal{B}_k^\ell$, there exist $B_0,\ldots,B_m \in \mathcal{B}_k$ with $B_0 = B$, $B_m = B^\ell$, $|B_{j-1}| < |B_j|$ (and hence $|B_{j-1}| \leq (2\gamma)^{-Q}|B_j|$), and $B_{j-1} \cap B_j \neq \emptyset$ for $j = 1,\ldots,m$. Suppose $x_0 \in B$, pick $x_j \in B_{j-1} \cap B_j$ for $1 \leq j \leq m$, and let x_{m+1} be the center of B^ℓ. Then for $0 \leq j \leq m$, x_j and x_{j+1} both lie in B_j, so

$$|x_j^{-1}x_{j+1}| < 2\gamma|B_j|^{1/Q} \leq (2\gamma)^{1-m+j}\ |B^\ell|^{1/Q}.$$

Therefore, by induction on the triangle inequality (1.8),

$$|x_0^{-1}x_{m+1}| = |x_0^{-1}x_1x_1^{-1}x_2 \cdots x_m^{-1}x_{m+1}|$$

$$\leq \gamma(\gamma(\cdots \gamma(|x_0^{-1}x_1| + |x_1^{-1}x_2|) + \cdots + |x_{m-1}^{-1}x_m|) + |x_m^{-1}x_{m+1}|)$$

$$\leq \Sigma_0^m \gamma^{m+1-j}\ |x_j^{-1}x_{j+1}|$$

$$< \Sigma_0^m \gamma^{m+1-j}(2\gamma)^{1-m+j}|B^\ell|^{1/Q}$$

$$\leq 4\gamma^2|B^\ell|^{1/Q},$$

and hence $B \subset B(4\gamma^2|B^\ell|^{1/Q}, x_{m+1}) \subset (B^\ell)^*$. #

Next, let

$$\theta_k^\ell = \Sigma_{B\in\mathcal{B}_k^\ell} F_B.$$

(7.20) LEMMA. *There exists* $C > 0$, *independent of* k *and* ℓ, *such that*

$$\|\theta_k^\ell\|_2^2 \leq C\Sigma_{B\in\mathcal{B}_k^\ell} S_B^2.$$

Proof: Let $B_1, B_2, \ldots$ be an enumeration of the elements of $\mathcal{B}_k^\ell$, ordered so that $|B_i| \geq |B_j|$ for $i \leq j$. This is possible by Lemmas 7.14 and 7.19, which guarantee that there are at most finitely many balls of a given size in $\mathcal{B}_k^\ell$. Also, for notational simplicity let us set $F_j = F_{B_j}$ and $S_j = S_{B_j}$. Then

$$\|\theta_k^\ell\|_2^2 = \Sigma_j \|F_j\|_2^2 + 2\,\mathrm{Re}\,\Sigma_{i<j} \int F_i \overline{F}_j.$$

We estimate the first sum on the right by Lemma 7.17:

$$\|F_j\|_2^2 = \int_{B_j^*} |F_j|^2 \leq C|B_j^*||B_j|^{-1}S_j^2 \leq C'S_j^2.$$

To estimate the cross terms we need only consider i,j such that $B_i^* \cap B_j^* \neq \emptyset$, for otherwise $F_i\overline{F}_j$ vanishes identically. If $i < j$ and

$B_i^* \cap B_j^* \neq \emptyset$, let x_j be the center of B_j and let P_{ij} be the left Taylor polynomial of F_i at x_j of homogeneous degree a. Then if $m = [a]$ and $b = \min\{b' \in \Delta : b' > a\}$, by Lemma 7.17 and the left-invariant version of the Taylor inequality (1.37), we have

$$\begin{aligned} \left|\int F_i \overline{F}_j\right| &= \left|\int (F_i - P_{ij})\overline{F}_j\right| \\ &\leq C \int_{B_j^*} \Sigma_{|I| \leq m+1, d(I) \geq b} |x_j^{-1}y|^{d(I)} \|X^I F_i\|_\infty \|F_j\|_\alpha dy \\ &\leq C\Sigma_{|I| \leq m+1, d(I) \geq b} |B_j^*|^{1+(d(I)/Q)} |B_i|^{-(1/2)-(d(I)/Q)} |B_j|^{-1/2} S_i S_j \\ &\leq C'(|B_j|/|B_i|)^{(1/2)+(b/Q)} S_i S_j \end{aligned}$$

since $|B_j|/|B_i| \leq 1$. Therefore, let

$$\alpha_{ij} = (|B_i|/|B_j|)^{(1/2)+(b/Q)} \quad \text{if } i < j \text{ and } B_i^* \cap B_j^* \neq \emptyset,$$

$$\alpha_{ij} = 0 \quad \text{otherwise.}$$

We must show that

$$\Sigma_{ij}\alpha_{ij}S_iS_j \leq C\Sigma_j S_j^2.$$

To do this it will suffice to show that for some $C > 0$,

$$(7.21) \qquad \Sigma_j \alpha_{ij} \leq C \text{ for all } i, \quad \Sigma_i \alpha_{ij} \leq C \text{ for all } j.$$

Indeed, by the Schwarz inequality we then have

$$\Sigma_i(\Sigma_j\alpha_{ij}S_j)^2 \leq \Sigma_i(\Sigma_j\alpha_{ij})(\Sigma_j\alpha_{ij}S_j^2)$$

$$\leq C\Sigma_{ij}\alpha_{ij}S_j^2 \leq C^2\Sigma_j S_j^2,$$

and hence

$$\Sigma_{ij}\alpha_{ij}S_iS_j \leq (\Sigma_i S_i^2)^{1/2}(\Sigma_i(\Sigma_j\alpha_{ij}S_j)^2)^{1/2}$$

$$\leq C\Sigma_j S_j^2.$$

But (7.21) follows easily from Lemma 7.14. In fact, for each i there are at most $N''(2\gamma)^{mQ}$ values of j such that $|B_j| = (2\gamma)^{-mQ}|B_i|$ and $B_i^* \cap B_j^* \neq \emptyset$, and for each j there are at most N' values of i such that $|B_i| = (2\gamma)^{mQ}|B_j|$ and $B_i^* \cap B_j^* \neq \emptyset$. Therefore,

$$\Sigma_j\alpha_{ij} \leq \Sigma_0^\infty N''(2\gamma)^{mQ}(2\gamma)^{-(mQ/2)-mb} = N''\Sigma_0^\infty(2\gamma)^{m[(Q/2)-b]},$$

where the sum converges since $b > a \geq Q/2$, and

$$\Sigma_i\alpha_{ij} \leq \Sigma_0^\infty N'(2\gamma)^{-m[(Q/2)+b]}. \quad \#$$

Proof of Theorem 7.11: We have

$$F = \Sigma_{B\in\mathcal{B}} F_B = \Sigma_{k\ell}\,\theta_k^\ell$$

where the sum converges in $\mathcal{S}'$, and we claim that the latter sum gives a decomposition of F into $(p,2,a)$-atoms. We first observe that by Lemma 7.19,

$$\operatorname{supp}\theta_k^\ell \subset \bigcup_{B\in\mathcal{B}_k^\ell} B^* \subset \tilde{B}^\ell,$$

where $\tilde{B}^\ell$ is the ball concentric with B^ℓ with radius $25\gamma^6$ times as big. Let

$$\lambda_k^\ell = \|\theta_k^\ell\|_2/|\tilde{B}^\ell|^{(1/2)-(1/p)}, \qquad a_k^\ell = \theta_k^\ell/\lambda_k^\ell.$$

Then a_k^ℓ is a $(p,2,a)$-atom, and $F = \Sigma_{k\ell}\lambda_k^\ell a_k^\ell$. Moreover, by Lemma 7.20 and Hölder's inequality,

$$\Sigma_{k\ell}(\lambda_k^\ell)^p \leq C\Sigma_{k\ell}(\Sigma_{B\in\mathcal{B}_k^\ell} S_B^2)^{p/2} |B^\ell|^{1-(p/2)}$$

$$\leq C\Sigma_k(\Sigma_\ell\Sigma_{B\in\mathcal{B}_k^\ell} S_B^2)^{p/2}(\Sigma_\ell|B^\ell|)^{1-(p/2)}.$$

But by Lemma 7.18,

$$\Sigma_\ell\Sigma_{B\in\mathcal{B}_k^\ell} S_B^2 = \Sigma_{B\in\mathcal{B}_k} S_B^2 \leq C(2\gamma)^{2k}|\Omega_k|.$$

Moreover, the B^ℓ's are disjoint unless they have the same size, so by Lemma 7.14,

$$\Sigma_\ell|B^\ell| \leq 2\Sigma_\ell|B^\ell \cap \Omega_k| \leq 2N|\Omega_k|.$$

Therefore,

$$\Sigma_{k\ell}(\lambda_k^\ell)^p \leq C\Sigma_k(2\gamma)^{kp}|\Omega_k|$$

$$= C\Sigma_k(2\gamma)^{k(p-1)}|\Omega_k|(2\gamma)^k$$

$$\leq C'\int_0^\infty p\lambda^{p-1}|\{x : S_\phi f(x) > \lambda\}|d\lambda$$

$$= C'\|S_\phi f\|_p^p.$$

This completes the proof. #

(7.22) COROLLARY. *Suppose* $0 < p \le 1$, $\phi^1,\dots,\phi^N,\psi^1,\dots,\psi^N \in \mathcal{S}$, $\int\phi^j = 0$ *for* $1 \le j \le N$, *and* $\Sigma_1^N \int_0^\infty \phi_t^j * \psi_t^j dt/t = \delta$ *in the sense of* (1.59'). *Suppose moreover that for* $1 \le j \le N$, $\int \psi^j P = 0$ *for all* $P \in \mathcal{P}_{[\bar{a}][\bar{a}+1](M+1)}$ *where* $M \in \mathbb{N}$ *is p-admissible and* $M \ge Q/2$. *If* $f \in \mathcal{S}'$ *vanishes weakly at infinity and* $S_{\phi^j} f \in L^p$ *for* $1 \le j \le N$, *then* $f \in H^p$, *and* $\rho^p(f) \le C_p \Sigma_1^N \|S_{\phi^j} f\|_p^p$.

Proof: By Theorem 1.62, there exist $\Phi^1,\dots,\Phi^{N'},\Psi^1,\dots,\Psi^{N'} \in \mathcal{S}$ which satisfy the hypotheses of Theorem 7.11, such that each Φ^k is of the form $\phi^j * \eta$ for some $\eta \in \mathcal{S}$ and some $j \in \{1,\dots,n\}$. By Corollary 7.6, $\|S_{\Phi^k} f\|_p \le C\|S_{\phi^j} f\|_p$, so we can apply Theorem 7.11. #

For the remainder of this chapter we assume that G is a stratified group, and we investigate the Lusin and Littlewood-Paley functions associated to the heat kernel h on G. For $j = 1,2,3,\dots$, let

$$\phi^{(j)}(x) = \partial_t^j h(x,t)\big|_{t=1} = (-\mathcal{L})^j h(x,1).$$

Then

$$(7.23)\qquad \int \phi^{(j)}(x)P(x)dx = \int h(x,1)(-\mathcal{L})^j P(x)dx = 0 \qquad \text{for } P \in \mathcal{P}_{2j-1}.$$

In particular, $\int \phi^{(j)} = 0$ for all j. If $f \in \mathcal{S}'$, we set

$$g_j f = g_{\phi^{(j)}} f, \qquad S_j f = S_{\phi^{(j)}} f.$$

Sometimes it is convenient to express $g_j f$ and $S_j f$ directly in terms of the heat semigroup H_t. For this purpose it is necessary to make the change of variable $t \to t^2$ to reflect the homogeneity of h. Indeed, we have

$$(7.24)\qquad \partial_t^j h(x,t) = t^{-(Q/2)-j}(\partial_t^j h)(x/\sqrt{t},1) = t^{-j}\phi_{\sqrt{t}}^{(j)}(x).$$

Hence,

$$(g_j f(x))^2 = \int_0^\infty |f*\phi_s^{(j)}(x)|^2 ds/s = \frac{1}{2}\int_0^\infty |f*\phi_{\sqrt{t}}^{(j)}(x)|^2 dt/t$$

$$= \frac{1}{2}\int_0^\infty |\partial_t^j H_t f(x)|^2 t^{2j-1} dt,$$

and

$$(S_j f(x))^2 = \int_0^\infty \int_{|x^{-1}y|<s} |f*\phi_s^{(j)}(y)|^2\, s^{-Q-1} dy ds$$

$$= \frac{1}{2}\int_0^\infty \int_{|x^{-1}y|^2<t} |f*\phi_{\sqrt{t}}^{(j)}(y)|^2\, t^{-(Q/2)-1} dy dt$$

$$= \frac{1}{2}\int_0^\infty \int_{|x^{-1}y|^2<t} |\partial_t^j H_t f(y)|^2\, t^{2j-(Q/2)-1} dy dt.$$

If $f \in H^p$ $(0 < p < \infty)$, by Theorems 7.7 and 7.8 we have $g_j f \in L^p$ and $S_j f \in L^p$ for all j. We now aim to obtain the converse of this statement. We shall have several occasions to use the following identity, which is valid since convolution with h commutes with L:

$$(7.25)\qquad \partial_t^{j+k} h(\cdot,t_1+t_2) = (-L)^{j+k} h(\cdot,t_1+t_2) = (-L)^{j+k}[h(\cdot,t_1)*h(\cdot,t_2)]$$

$$= (-L)^j h(\cdot,t_1)*(-L)^k h(\cdot,t_2) = \partial_t^j h(\cdot,t_1)*\partial_t^k h(\cdot,t_2).$$

(7.26) PROPOSITION. For all positive integers j and k,

$$\frac{(-1)^{j+k}2^{j+k+1}}{(j+k-1)!}\int_0^\infty \phi_s^{(j)}*\phi_s^{(k)}\,ds/s = \delta.$$

Proof: To begin with, we have

$$\delta = -h(\cdot,t)\big|_0^\infty = -\int_0^\infty \partial_t h(\cdot,t)\,dt.$$

If we insert a factor of

$$1 = (d/dt)^{j+k-1}t^{j+k-1}/(j+k-1)!$$

into this integral and integrate by parts $j+k-1$ times, we obtain

$$\delta = \frac{(-1)^{j+k}}{(j+k-1)!}\int_0^\infty t^{j+k-1}\partial_t^{j+k}h(\cdot,t)\,dt,$$

Thus, by (7.24), (7.25), and the substitution $t = s^2$,

$$\delta = \frac{(-1)^{j+k}}{(j+k-1)!}\int_0^\infty t^{j+k-1}\partial_t^j h(\cdot,t/2)*\partial_t^k h(\cdot,t/2)\,dt$$

$$= \frac{(-1)^{j+k}}{(j+k-1)!}\int_0^\infty t^{j+k-1}(t/2)^{-j-k}\phi_{\sqrt{t}}^{(j)}*\phi_{\sqrt{t}}^{(k)}\,dt$$

$$= \frac{(-1)^{j+k}2^{j+k}}{(j+k-1)!}\int_0^\infty \phi_{\sqrt{t}}^{(j)}*\phi_{\sqrt{t}}^{(k)}\,dt/t$$

$$= \frac{(-1)^{j+k}2^{j+k+1}}{(j+k-1)!}\int_0^\infty \phi_s^{(j)}*\phi_s^{(k)}\,ds/s. \quad \#$$

(7.27) COROLLARY. <u>If</u> $f \in S'$ <u>vanishes weakly at infinity and</u> $S_j f \in L^p$ $(0 < p < \infty)$ <u>for some</u> $j \geq 1$, <u>then</u> $f \in H^p$.

<u>Proof</u>: This follows from Theorem 7.10 and Corollary 7.22 in view of (7.23) and Proposition 7.26, since $\phi^{(k)}$ will have as many vanishing moments as we please if we take k sufficiently large. #

(7.28) THEOREM. <u>If</u> $f \in S'$ <u>vanishes weakly at infinity and</u> $g_j f \in L^p$ $(0 < p < \infty)$ <u>for some</u> $j \geq 1$, <u>then</u> $f \in H^p$.

<u>Proof</u>: If $p > 1$ this follows from Theorem 7.10 and Proposition 7.26. For $p \leq 1$, our strategy will be to show that if $g_j f \in L^p$ then $S_{j+1} f \in L^p$, which yields the desired result by virtue of Corollary 7.27.

Let $X = L^2((0,\infty), t^{2j-1}dt)$, and define $F : G \to X$ and $V : G \times (0,\infty) \to X$ by

$$F(x)(t) = \partial_t^j H_t f(x), \qquad V(x,s) = H_s F(x).$$

(Thus, by (7.24), $V(x,s)(t) = \partial_t^j H_{t+s} f(x)$.) Then

$$\|V(x,s)\|_X^2 = \int_0^\infty |\partial_t^j H_{t+s} f(x)|^2 t^{2j-1} dt = \int_s^\infty |\partial_t^j H_t f(x)|^2 (t-s)^{2j-1} dt,$$

and hence

$$\sup_{s>0} \|V(x,s)\|_X^2 = \int_0^\infty |\partial_t^j H_t f(x)|^2 t^{2j-1} dt = (g_j f(x))^2.$$

In other words,

$$M_\phi^0 F = g_j f \in L^p, \qquad \text{where} \quad \phi = h(\cdot,1).$$

Since ϕ is a commutative approximate identity, by (the vector-valued analogue of) Corollary 4.17 we have $F \in H^p_X$, and hence by Theorems 7.8 and 7.1 (with $\beta = 1$, $\alpha = 2$), $S_1^2 F \in L^p$. (The superscript 2 here refers to the aperture of the cone defining $S_1^2 F$.) But by (7.25),

$$[S_1^2 F(x)]^2 = \int_0^\infty \int_{|x^{-1}y|^2<2t} \|\partial_t H_t F(y)\|_X^2 \, t^{-(Q/2)-1} dydt$$

$$= \int_0^\infty \int_{|x^{-1}y|^2<2t} \int_0^\infty |\partial_t^{j+1} H_{t+s} f(y)|^2 \, s^{2j-1} t^{-(Q/2)-1} dsdydt.$$

We integrate first with respect to y, then t, then s, and replace the variable t by $\sigma = t+s$, obtaining

$$[S_1^2 F(x)]^2 = \int_0^\infty \int_s^\infty \int_{|x^{-1}y|^2<2(\sigma-s)} |\partial_\sigma^{j+1} H_\sigma f(y)|^2 \, s^{2j-1} (\sigma-s)^{-(Q/2)+1} dyd\sigma ds.$$

Now we interchange the s and σ integrations and shrink the interval of integration in s:

$$\int_0^\infty \int_s^\infty (\cdots) d\sigma ds = \int_0^\infty \int_0^\sigma (\cdots) dsd\sigma \geq \int_0^\infty \int_0^{\sigma/2} (\cdots) dsd\sigma.$$

Since $s \leq \sigma/2$ implies $2(\sigma-s) \geq \sigma \geq \sigma-s \geq \sigma/2$, then,

$$[S_1^2 F(x)]^2 \geq \int_0^\infty \int_0^{\sigma/2} \int_{|x^{-1}y|^2<\sigma} |\partial_\sigma^{j+1} H_\sigma f(y)|^2 \, s^{2j-1} \sigma^{-(Q/2)+1} dydsd\sigma$$

$$= (2j)^{-1} \int_0^\infty \int_{|x^{-1}y|^2<\sigma} |\partial_\sigma^{j+1} H_\sigma f(y)|^2 (\sigma/2)^{2j} \sigma^{-(Q/2)+1} dyd\sigma$$

$$= j^{-1} 2^{-2j-1} [S_{j+1} f(x)]^2.$$

Therefore $S_{j+1} f \in L^p$, as desired. #

Notes and References

The classical Lusin and Littlewood-Paley functions (for the unit disc) are in Zygmund [1]. The Lusin and Littlewood-Paley functions on $\mathbb{R}^n$ were initially defined in terms of Poisson integrals. In this setting, the characterization of L^p $(1 < p < \infty)$ by the S and g functions is due to Stein [1], and the characterization of H^p $(p \leq 1)$ by these functions is due to C. Fefferman and Stein [1]; see also Calderón [2] and Segovia [1]. An analogue on the Heisenberg group has been obtained by Geller [2]. For more general Lusin and Littlewood-Paley functions on $\mathbb{R}^n$, most of the results in this chapter were first proved by Calderón and Torchinsky [1]. See also Stein [3] for some further generalization of the theory for $p > 1$.

Our proof of Theorem 7.11 is inspired by some ideas of Chang and R. Fefferman [1].

CHAPTER 8

Boundary Value Problems

Classical H^p theory deals with the boundary behavior of solutions of certain partial differential equations such as the Cauchy-Riemann equations or Laplace's equation. This chapter is devoted to some results which link our real-variable theory to the classical framework. In Section A we show that if G is stratified, H^p can be characterized as the set of boundary distributions of temperatures (i.e., solutions of the heat equation) on $G \times (0,\infty)$ whose maximal functions are in L^p. In Section B we show that for the purposes of H^p theory the class $\mathcal{S}$ of test functions can be enlarged to include certain Poisson-type kernels, thereby making the connection with harmonic functions. Finally, in Section C we investigate the H^p behavior of Poisson integrals on symmetric spaces. The Poisson integral on a symmetric space is of the type treated in Section B only if the symmetric space is of rank one. The study of the general higher rank case (more precisely the questions related to "restricted convergence" of Poisson integrals) has a long history. Here the main feature of the problem is that we are dealing with approximate identities fashioned out of ϕ which are only slowly decreasing at infinity in some directions. This makes the analysis quite delicate, as can already be seen in the early work of Marcinkiewicz and Zygmund, (see Zygmund [1], Chapter 17), where $\mathbb{R}^n$ is considered as the (distinguished) boundary of the product of n upper half-planes.

A. Temperatures on Stratified Groups

In this section we assume that G is a stratified group, and we study the boundary behavior of solutions of the heat equation $(\partial_t + L)u = 0$ on $G \times (0,\infty)$. In addition to the facts about the heat kernel presented in Chapter 1, Section G, we shall need the following maximum principle for $\partial_t + L$, due to Bony [1]:

(8.1) PROPOSITION. *Let Ω be an open set in $G \times \mathbb{R}$, and let $u \in C^\infty(\Omega)$ be a real-valued solution of $(\partial_t + L)u = 0$ on Ω. Suppose that u attains its supremum or infimum on* Ω *at $(x_0,t_0) \in \Omega$. Then if $\gamma : [0,1] \to \Omega$ is any smooth curve such that $\gamma(0) = (x_0,t_0)$ and $\gamma'(\tau)\cdot\partial_t \leq 0$ for $0 \leq \tau \leq 1$ (that is, $\gamma'(\tau) = \Sigma a_j(\gamma(\tau))X_j + b(\gamma(\tau))\partial_t$ where $b \leq 0$), then $u(\gamma(\tau)) = u(x_0,t_0)$ for $0 \leq \tau \leq 1$.*

(8.2) COROLLARY. *The heat kernel $h(x,t)$ is strictly positive for $t > 0$.*

Proof: We know that $h(x,t) \geq 0$. If $h(x_0,t_0) = 0$ with $t_0 > 0$ we would have $h(x,t) = 0$ for all $x \in G$ and $t < t_0$, which is false since $\int h(x,t)dx = 1$. #

(8.3) COROLLARY. *Suppose Ω is a bounded open set in $G \times \mathbb{R}$ and u is a (complex-valued) solution of $(\partial_t + L)u = 0$ on Ω which is continuous on $\overline{\Omega}$. Then $\sup_\Omega |u(x,t)| = \sup_{\partial\Omega} |u(x,t)|$.*

Proof: Suppose $|u(x_0,t_0)| = \sup_\Omega |u(x,t)|$ where $(x_0,t_0) \in \Omega$. By multiplying u be a constant of modulus 1 we may assume that $u(x_0,t_0) > 0$.

For any $X \in \mathfrak{g}$, the curve $\gamma(\tau) = (x_0 \exp(\tau X), t_0)$ eventually intersects $\partial\Omega$, and $\operatorname{Re}(u)$ is constant on this curve, so $|u(x_0,t_0)| \leq \sup_{\partial\Omega}|u(x,t)|$. The reverse inequality is trivial. #

If u is any continuous function on $G \times (0,\infty)$, we define the maximal function u^* on G by

$$u^*(x) = \sup_{|x^{-1}y|^2<t<\infty} |u(y,t)|.$$

Note that we are using the inequality $|x^{-1}y|^2<t$ rather than $|x^{-1}y|<t$ to determine the "non-tangential cone." The reason for this is to make the definition fit with the homogeneity of the heat kernel. Indeed, suppose $u(x,t) = H_t f(x)$ for some $f \in \mathcal{S}'$, and let $\phi(x) = h(x,1)$. Then $h(x,t) = \phi_{\sqrt{t}}(x)$, so

$$u^*(x) = \sup_{|x^{-1}y|^2<t} |f*\phi_{\sqrt{t}}(y)| = \sup_{|x^{-1}y|<s} |f*\phi_s(y)| = M_\phi f(x).$$

Combining this observation with Corollary 4.11, we have:

(8.4) PROPOSITION. _Suppose_ $0 < p \leq \infty$ _and_ $f \in \mathcal{S}'$. _Then_ $f \in H^p$ _if and only if_ $u^* \in L^p$, _where_ $u(x,t) = H_t f(x)$.

What remains to be settled is the following question: if $(\partial_t + \mathcal{L})u = 0$ on $G \times (0,\infty)$ and $u^* \in L^p$, is $u(x,t)$ equal to $H_t f(x)$ for some $f \in H^p$? We shall show that the answer is affirmative.

(8.5) Lemma. _If_ u _is continuous on_ $G \times (0,\infty)$ _and_ $u^* \in L^p$ _then_

$$|u(x,t)| \leq \|u^*\|_p \, t^{-Q/2p}.$$

<u>Proof</u>: Since $|u(x,t)| \leq u^*(y)$ whenever $|x^{-1}y|^2 < t$, we have

$$|u(x,t)|^p \leq |B(\sqrt{t},x)|^{-1} \int_{B(\sqrt{t},x)} u^*(y)^p dy \leq t^{-Q/2} \|u^*\|_p^p. \quad \#$$

(8.6) LEMMA. <u>Suppose</u> u <u>is continuous on</u> $G \times (0,\infty)$ <u>and</u> $u^* \in L^p$ $(0 < p < \infty)$. <u>For any</u> $\varepsilon > 0$ <u>and</u> $\delta > 0$ <u>there exist</u> $T > 0$ <u>and</u> $R > 0$ <u>such that</u> $|u(x,t)| < \delta$ <u>when</u> $t > T$ <u>or when</u> $\varepsilon \leq t \leq T$ <u>and</u> $|x| > R$.

<u>Proof</u>: By Lemma 8.5 we have $|u(x,t)| < \delta$ when $t < T$ provided $T \geq (\delta/\|u^*\|_p)^{-2p/Q}$. Also, as in the proof of Lemma 8.5 we have

$$|u(x,t)|^p \leq |B(\sqrt{t},x)|^{-1} \int_{B(\sqrt{t},x)} u^*(y)^p dy = (u^*)^p * \chi_{\sqrt{t}}(x)$$

where χ is the characteristic function of $B(1,0)$. Since $(u^*)^p \in L^1$ we can write $(u^*)^p = f+g$ where f has compact support and $\|g\|_1 < \varepsilon^{Q/2}\delta^p$. Then there exists $R > 0$ such that $f*\chi_{\sqrt{t}}$ is supported in $B(R,0)$ for $t \leq T$. Also, if $t \geq \varepsilon$,

$$\|g*\chi_{\sqrt{t}}\|_\infty \leq \|g\|_1 \|\chi_{\sqrt{t}}\|_\infty < \varepsilon^{Q/2}\delta^p t^{-Q/2} \leq \delta^p.$$

Thus $|u(x,t)|^p < \delta^p$ if $\varepsilon \leq t \leq T$ and $|x| > R$. #

(8.7) LEMMA. <u>Suppose</u> $f \in L^q \cap C_0$ <u>where</u> $1 \leq q < \infty$, <u>and let</u> $u(x,t) = H_t f(x)$. <u>For any</u> $\delta > 0$ <u>there exist</u> $T,R > 0$ <u>such that</u> $|u(x,t)| < \delta$ <u>when</u> $t > T$ <u>or when</u> $0 \leq t \leq T$ <u>and</u> $|x| > R$.

<u>Proof</u>: We have $|H_t f(x)| \leq \|f\|_q \|h(\cdot,t)\|_{q'} \leq C\|f\|_q t^{-Q/2q}$, so $|u(x,t)| < \delta$ for $t > T$ provided $T \geq (\delta/C\|f\|_q)^{-2q/Q}$. Also, since $f \in C_0$

we can write $f = f_1+f_2$ where f_1 has compact support and $\|f_2\|_\infty < \delta/2$, so that $|H_t f_2(x)| \leq \|f_2\|_\infty \|h(\cdot,t)\|_1 < \delta/2$. Finally, we can pick $r > 0$ so large that $\int_{|x|>r} h(x,t)dx < \delta/2\|f_1\|_\infty$ for $t \leq T$. Then

$$H_t f_1(x) = \left[\int_{|y|\leq r} + \int_{|y|>r}\right] f_1(xy^{-1})h(y,t)dy.$$

The first term on the right vanishes if $|x|$ is sufficiently large, and the second term is less than $\delta/2$ for $t \leq T$. #

(8.8) THEOREM. Suppose u satisfies $(\partial_t+L)u = 0$ on $G\times(0,\infty)$ and $u^* \in L^p$ $(0 < p < \infty)$. Then there exists $f \in S'$ such that $u(\cdot,\varepsilon) \to f$ in S' as $\varepsilon \to 0$, and $u(x,t) = H_t f(x)$. Consequently, H^p is precisely the set of boundary distributions of temperatures u on $G\times(0,\infty)$ such that $u^* \in L^p$.

Proof: Given $\phi \in S$, let $F(t) = \int u(x,t)\phi(x)dx$. (The integral converges by Lemma 8.5.) We must show that $\lim_{t\to 0} F(t)$ exists and is bounded by $C\|\phi\|_{(N)}$ for some $C > 0$ and $N \in \mathbb{N}$ independent of ϕ. Observe that for any $k \in \mathbb{N}$,

$$F^{(k)}(t) = \int \partial_t^k u(x,t)\phi(x)dx = \int (-L)^k u(x,t)\phi(x)dx = \int u(x,t)(-L)^k\phi(x)dx,$$

and thus by Lemma 8.5,

$$|F^{(k)}(t)| \leq \|u(\cdot,t)\|_\infty \int |L^k\phi(x)|dx \leq C\|\phi\|_{(2k)}\, t^{-Q/2p}. \tag{8.9}$$

In particular, $F^{(k)}(t) \to 0$ as $t \to \infty$ for all k, so for $k \geq 1$,

$$F^{(k-1)}(t) = -\int_t^\infty F^{(k)}(s)ds. \tag{8.10}$$

Let $N = [Q/2p]+1$. If $Q/2p$ is not an integer, taking k successively equal to $N, N-1, \ldots, 2$ in (8.10) and using (8.9) to estimate the integrand in (8.10) for $k = N$, we find that

$$|F'(t)| \leq C'\|\phi\|_{(2N)}\, t^{-(Q/2p)+N-1}.$$

If $Q/2p$ is an integer, the same argument yields $|F''(t)| \leq C'\|\phi\|_{(2N)}\, t^{-1}$, so

$$|F'(t)| \leq |F'(1)| + \left|\int_t^1 F''(s)ds\right| \leq C''\|\phi\|_{(2N)}\,(1+|\log t|) \qquad (t \leq 1).$$

In either case, F' is integrable on $(0,1)$, so

$$F(0) = F(1) - \lim_{t\to 0}\int_t^1 F'(s)ds$$

exists and is bounded by $C\|\phi\|_{(2N)}$. This completes the proof of the existence of $f \in S'$ such that $\lim_{\varepsilon\to 0} u(\cdot,\varepsilon) = f$.

Next, given $\varepsilon > 0$, let $v_\varepsilon(x,t) = u(x,t+\varepsilon)$ and $w_\varepsilon(x,t) = H_t(u(\cdot,\varepsilon))(x)$ Since $u^* \in L^p$ we have $u(\cdot,\varepsilon) \in L^p$, and $u(\cdot,\varepsilon) \in C_0$ by Lemma 8.6, so that we can apply Lemma 8.7 with $q = \max(p,1)$. Applying Lemma 8.6 to u and Lemma 8.7 to w_ε, we see that for any $\delta > 0$ there exist $T, R > 0$ such that $|v_\varepsilon(x,t) - w_\varepsilon(x,t)| < 2\delta$ when $t > T$ or when $0 \leq t \leq T$ and $|x| > R$. Moreover, v_ε and w_ε are continuous on $G \times [0,\infty)$ and $v_\varepsilon(x,0) = w_\varepsilon(x,0) = u(x,\varepsilon)$, so it follows from Corollary 8.3 that $|v_\varepsilon - w_\varepsilon| < 2$ everywhere. Since δ is arbitrary, $v_\varepsilon = w_\varepsilon$. But then

$$u(x,t) = \lim_{\varepsilon\to 0} v_\varepsilon(x,t) = \lim_{\varepsilon\to 0} w_\varepsilon(x,t) = \lim_{\varepsilon\to 0} H_t(u(\cdot,\varepsilon))(x)$$

$$= H_t f(x),$$

and we are done. #

B. Poisson Integrals on Stratified Groups

The H^p spaces on $\mathbb{R}^n$ were originally defined in terms of Poisson integrals. Since the Poisson kernel does not belong to $\mathcal{S}$, this approach to H^p does not fall within the scope of our theory so far. However, we shall now show that H^p can be characterized by versions of the Poisson integral in many cases. We begin by introducing a class of test functions which will include the Poisson kernels we have in mind.

We adopt the following notation throughout this section: if ϕ is a function on G, Φ will denote the function on $G \times (0,\infty)$ defined by $\Phi(x,t) = \phi_t(x) = t^{-Q}\phi(x/t)$. We observe that Φ is jointly homogeneous in x and t of degree $-Q$: $\Phi(rx,rt) = r^{-Q}\Phi(x,t)$.

Let $\mathcal{R}$ denote the class of all C^∞ functions ϕ on G such that (i) Φ and all its derivatives in x and t extend continuously to $(G \times [0,\infty))\backslash\{(0,0)\}$, and (ii) $\Phi(x,0) = 0$ for $x \neq 0$. Clearly $\mathcal{S} \subset \mathcal{R}$, since if $\phi(x)$ vanishes to infinite order as $x \to \infty$ then $\phi(x/t)$ vanishes to infinite order as $t \to 0$, uniformly for x in any subset of G not containing the origin. The other examples of functions in $\mathcal{R}$ in which we are interested are the following.

(1) Let G be a stratified group with sub-Laplacian $\mathcal{L}$. Then there is a unique $\phi \in \mathcal{R}$ with the following property: if f is bounded and continuous on G, the function

$$u(x,t) = f*\phi_t(x) = \int f(y)\Phi(y^{-1}x,t)dy$$

solves the Dirichlet problem

$$\mathcal{L}u - \partial_t^2 u = 0 \quad \text{on} \quad G \times (0,\infty), \qquad u(x,0) = f(x).$$

Φ is called the Poisson kernel for G; for its construction, see Folland [3].

(2) Let $G = H_n$ be the $(2n+1)$-dimensional Heisenberg group defined in Chapter 1, Section A, with the standard dilations. The Poisson-Szegö kernel on H_n is the function P on $H_n \times (0,\infty)$ defined by

$$P((z,t),\rho) = \frac{2^{n-1}n!}{\pi^n} \frac{\rho^n}{((\rho+|z|^2)^2+t^2)^{n+1}}.$$

P has the following interpretation: H_n can be canonically identified with the boundary of the domain

$$D_{n+1} = \{w \in \mathbb{C}^{n+1} : \operatorname{Im} w_{n+1} > \Sigma_1^n |w_j|^2\}$$

via the map $(z,t) \to (z,t+i|z|^2)$. If f is bounded and continuous on H_n, the function u on D_{n+1} defined by

$$u(z_1,\ldots,z_n,t+i(\rho+|z|^2)) = \int f(z',t')P((z',t')^{-1}(z,t),\rho)d(z',t')$$

solves the Dirichlet problem

$$\Delta_B u = 0 \quad \text{on} \quad D_{n+1}, \qquad u(z,t+i|z|^2) = f(z,t)$$

where Δ_B is the Laplace-Beltrami operator associated to the Bergman metric on D_{n+1}. (See, e.g., Korányi [1]. P is actually a special case of the Poisson kernels we consider in Section C.) Let $\phi(z,t) = P((z,t),1)$: then $\phi \in \mathcal{R}$, for $\Phi((z,t),r) = P((z,t),r^2)$.

(3) A formula very similar in appearance to the above Poisson-Szegö kernel holds for the Poisson kernel associated with any symmetric space of rank 1. (See Helgason [1], p. 59). We omit any further details but state here that the resulting ϕ is also in $\mathcal{R}$.

We return now to the general situation.

(8.11) PROPOSITION. If $\phi \in \mathcal{R}$, then:

(a) $|Y^I \partial_t^j \Phi(x,t)| \leq C_{I_j}(t + |x|)^{-(Q+d(I)+j)}$,

(b) $|Y^I \phi(x)| \leq C_I(1 + |x|)^{-(Q+d(I)+1)}$,

(c) $\phi \in L^1$.

Proof: (a) follows from the homogeneity of Φ and its smoothness as $t \to 0$: in fact, we can take

$$C_{I_j} = \sup_{t+|x|=1} |Y^I \partial_t^j \Phi(x,t)|.$$

Next, we observe that since $\Phi(x,0) = 0$ we also have $Y^I \Phi(x,0) = 0$, and since $Y^I \Phi(x,t)$ is smooth as $t \to 0$ for $x \neq 0$,

$$\sup_{|y|=1, 0 \leq t \leq 1} |Y^I \Phi(y,t)|/t = C_I < \infty.$$

However,

$$Y^I \Phi(y,t) = t^{-Q-d(I)}(Y^I \phi)(y/t).$$

Hence if $|x| \geq 1$ and we set $y = x/|x|$ and $t = 1/|x|$, we obtain

$$|Y^I \phi(x)| = |x|^{-Q-d(I)} |Y^I \Phi(y,t)| \leq C_I |x|^{-Q-d(I)-1},$$

which implies (b). Finally, if we take $I = 0$ in (b) and integrate, we obtain (c). #

The estimate (a) in Proposition 8.11 suggests a natural topology for $\mathcal{R}$: namely, we define the norms $\| \ \|_{(I,j)}$ on $\mathcal{R}$ by

$$\|\phi\|_{(I,j)} = \sup_{x\in G, t\geq 0} (t + |x|)^{Q+d(I)+j} |Y^I \partial_t^j \phi(x,t)|.$$

$\mathcal{R}$ is easily seen to be a Fréchet space when equipped with this family of norms. (It is not, however, the full space of functions for which these norms are finite, since the condition $\Phi(x,0) = 0$ imposes additional restrictions.)

We shall denote by $\mathcal{R}'$ the space of continuous linear functionals on $\mathcal{R}$. The natural restriction map from $\mathcal{R}'$ to $\mathcal{S}'$ is not injective, since $\mathcal{S}$ is not dense in $\mathcal{R}$. (For example, if $x_0 \neq 0 \in G$ and $j \geq 1$, the functional $\phi \to \partial_t^j \Phi(x_0,t)|_{t=0}$ is an element of $\mathcal{R}'$ which annihilates $\mathcal{S}$.) However, since $\mathcal{R} \subset L^1 \cap L^\infty$, functions in L^q $(1 \leq q \leq \infty)$ have an obvious interpretation as elements of $\mathcal{R}'$.

(8.12) <u>Remark</u>: The proof of Proposition 8.11(b) shows that

$$|Y^I \phi(x)| \leq C(\|\phi\|_{(I,0)} + \|\phi\|_{(I,1)})(1 + |x|)^{-Q-d(I)-1}$$

where C is independent of ϕ.

(8.13) THEOREM. (a) <u>Suppose</u> $0 < p \leq 1$ <u>and</u> $a = \max\{a' \in \Delta : a' \leq Q(p^{-1}-$ <u>There exist</u> $N \in \mathbb{N}$ <u>and</u> $C > 0$ <u>such that for all</u> (p,∞,a)-<u>atoms</u> f <u>we have</u> $\|Mf\|_p^p \leq C$, <u>where</u>

$$Mf(x) = \sup\{M_\phi f(x) : \phi \in \mathcal{R}, \|\phi\|_{(0,1)} \leq 1, \text{ and } \|\phi\|_{(I,0)} \leq 1 \text{ for } |I| \leq N\}.$$

(b) *If* $0 < p \leq \infty$, *the natural inclusion* $(H^p \cap L^\infty) \subset \mathcal{R}'$ *extends to a continuous injection of* H^p *into* $\mathcal{R}'$. *Moreover, there exists* $C > 0$ *such that* $\|Mf\|_p^p \leq C\rho^p(f)$ *for all* $f \in H^p$, *where* Mf *is as above (with* $N = 0$ *when* $p > 1$).

Proof: The proof of (a) is essentially the same as the proof of Theorem 2.9: we assume that f is an atom associated to $B(r,0)$, and we set $b = \min\{b' \in \Delta : b' > a\}$, $N =$ the smallest integer $\geq b$, and $\tilde{B} = B(2\gamma\beta^N r, 0)$. We estimate the integral of $(Mf)^p$ over $\tilde{B}$ by using the maximal theorem (2.4) (which is applicable in view of Remark 8.12), and we estimate the integral over $\tilde{B}^c$ by subtracting off a Taylor polynomial from ϕ. The estimate for the remainder which we need is the following:

If $\phi \in \mathcal{R}$ and $x \in G$, let P_x be the right Taylor polynomial of ϕ at x of homogeneous degree a. If $|x| \geq 2\gamma\beta^N|y|$ then

$$|\phi(yx) - P_x(y)| \leq C \sup_{|I| \leq N} \|\phi\|_{(I,0)} |y|^b |x|^{-Q-b}.$$

Let us prove this. By the Taylor inequality (1.37),

$$|\phi(yx) - P_x(y)| \leq C\Sigma_{|I| \leq N, d(I) \geq b} |y|^{d(I)} \sup_{|z| \leq \beta^N |y|} |Y^I \phi(zx)|$$

$$\leq C\Sigma_{|I| \leq N, d(I) \geq b} |y|^{d(I)} \sup_{|z| \leq \beta^N |y|} \|\phi\|_{(I,0)} (1 + |zx|)^{-Q-d(I)}.$$

But if $|x| \geq 2\gamma\beta^N|y|$ and $|z| \leq \beta^N|y|$ then $|zx| \geq |x|/2$, so

$$|\phi(yx) - P_x(y)| \leq C|y|^b|x|^{-Q-b} \Sigma_{|I| \leq N, d(I) \geq b} \|\phi\|_{(I,0)} (|y|/|x|)^{d(I)-b}$$

$$\leq C|y|^b|x|^{-Q-b} \Sigma_{|I| \leq N, d(I) \geq b} \|\phi\|_{(I,0)} (2\gamma\beta^N)^{b-d(I)}$$

The desired estimate is now immediate, and the rest of the argument proceeds as before. In the proof of Theorem 2.9 we estimated the radial maximal functions $M_\phi^0 f$, but a routine modification of the argument yields the same estimates for $M_\phi f$; alternatively, one can prove the obvious analogue of Proposition 2.8. Details are left to the reader.

If $p \leq 1$, part (b) follows from part (a) and the atomic decomposition theorem by the argument used to prove Proposition 2.15. If $p > 1$, part (b) follows from the maximal theorem (2.4) and Remark 8.12. #

In proving Theorem 8.13 we did not use all the estimates in Proposition 8.11. The full force of the smoothness as $t \to 0$, however, will be used in the following arguments, which lead up to a generalization of Corollary 4.17.

(8.14) LEMMA. *There is a* C^∞ *function* ζ *on* $[1,\infty)$ *which is rapidly decreasing* (i.e., $|\zeta(s)| = O(s^{-N})$ *for all* N), *such that* $\int_1^\infty \zeta(s)ds = 1$ *and* $\int_1^\infty s^k \zeta(s)ds = 0$ *for all positive integers* k.

Proof: Let $\omega = e^{-i\pi/4}$, and consider the function $f(z) = \exp(-\omega(z-1)^{1/}$ on the complex plane cut along the ray $[1,\infty)$, where $0 < \arg(z-1) < 2\pi$. Let γ be the contour which goes from ∞ to 1 along the lower edge of the cut, makes an infinitesmial half-loop around 1, and returns to ∞ along the upper edge of the cut. Since f is rapidly decreasing at ∞, by Cauchy's theorem we have

$$\frac{1}{2\pi i}\int_\gamma f(z)\,\frac{dz}{z} = f(0) = e^{-1}; \qquad \int_\gamma z^k f(z)\,\frac{dz}{z} = 0 \quad \text{for } k \geq 1.$$

From this it is easily verified that we may take

$$\zeta(s) = \frac{e}{\pi s}\,\mathrm{Im}\,\exp(-\omega(s-1)^{1/4}) \qquad (\arg s = 0). \quad \#$$

(8.15) LEMMA. <u>Suppose</u> $\phi \in \mathcal{R}$ <u>and</u> ζ <u>is as in Lemma</u> 8.14. <u>If</u> $\psi(x) = \int_1^\infty \zeta(s)\Phi(x,s)ds$ <u>then</u> $\psi \in \mathcal{S}$.

<u>Proof</u>: Clearly differentiation under the integral sign is permissible, so that $\psi \in C^\infty$. We must show that $Y^I\psi$ is rapidly decreasing for all I. However, for any $N \in \mathbb{N}$, by the smoothness properties of Φ there exists $C_N > 0$ such that

$$\sup_{|y|=1} |Y^I\Phi(y,t) - \Sigma_0^{N-1}\partial_t^k Y^I\Phi(y,0)t^k/k!| \leq C_N t^N.$$

We observe that since $\Phi(y,0) = 0$ the term with $k = 0$ in the Taylor polynomial vanishes. Hence if $x \neq 0$ and we set $y = x/|x|$, $t = s/|x|$ $(s \geq 1)$, we obtain

$$Y^I\Phi(x,s) = |x|^{-Q-d(I)}\, Y^I\Phi(y,t)$$

$$= |x|^{-Q-d(I)}[\Sigma_1^{N-1}\partial_t^k Y^I\Phi(x/|x|,0)s^k/(k!\,|x|^k) + R_N(x,s)]$$

where $|R_N(x,s)| \leq C_N(s/|x|)^N$. But from this it follows that

$$Y^I\psi(x) = |x|^{-Q-d(I)} \int_1^\infty R_N(x,s)\zeta(s)ds = O(|x|^{-Q-d(I)-N}),$$

which completes the proof. #

As in Chapter 4, a function $\phi \in \mathcal{R}$ will be called a <u>commutative approximate identity</u> if $\int \phi = 1$ and $\phi_s * \phi_t = \phi_t * \phi_s$ for all $s,t > 0$. The Poisson kernels associated to sub-Laplacians on stratified groups and the Poisson-Szegö kernel on H_n discussed above (or rather the ϕ's for which they are the corresponding Φ's) are commutative approximate identities. Indeed, the condition $\int \phi = 1$, in both cases, falls out of the fact that $f*\phi_t \to f$ as $t \to 0$, by Proposition 1.20. The Poisson kernels associated to sub-Laplacians have the semigroup property $\phi_s * \phi_t = \phi_{s+t}$ (cf. Folland [3]), while the Poisson-Szegö kernel is commutative because it is polyradial (cf. Proposition 4.28).

(8.16) THEOREM. Let $\phi \in \mathcal{R}$ be a commutative approximate identity. If $f \in \mathcal{R}'$ and $M_\phi^0 f \in L^p$, then the restriction of f to S is in H^p.

Proof: Let $\psi(x) = \int_1^\infty \zeta(s)\Phi(x,s)ds$ as in Lemma 8.15. Then $\psi \in S$,

$$\int \psi(x)dx = \int_1^\infty \int_G \zeta(s)\Phi(x,s)dxds = \int_1^\infty \zeta(s)ds = 1,$$

and it is easily checked that $\psi_s * \psi_t = \psi_t * \psi_s$ since the same is true of ϕ. Moreover,

$$M_\psi^0 f(x) = \sup_{t>0} \left| \int_1^\infty f*\phi_{st}(x)\zeta(s)ds \right|$$

$$\leq \left(\int_1^\infty |\zeta(s)|ds \right) M_\phi^0 f(x).$$

Hence ψ is a commutative approximate identity and $M_\psi^0 f \in L^p$, whence $f|S \in H^p$ by Corollary 4.17. #

Remark: When G is stratified, the analogue of Theorem 8.8 for solutions of $(L-\partial_t^2)u = 0$ on $G \times (0,\infty)$ is true, with essentially the same proof. The relevant maximum principle for $L-\partial_t^2$ is in Bony [1].

C. Poisson Integrals on Symmetric Spaces

In this section we prove a result concerning the boundary behavior of harmonic functions on symmetric spaces of noncompact type. We shall sketch the relevant background very briefly; a more complete exposition of the needed material can be found in Korányi [2].

In order to conform with standard usage, we change our notation for groups. G will now denote a noncompact semisimple Lie group with finite center, with maximal compact subgroup K. Let $G = KAN$ be the Iwasawa decomposition of G. If $g \in G$, the K,A,N components of g will be denoted by $\kappa(g)$, $\exp H(g)$, $\nu(g)$. $\overline{N}$ will denote the image of N under the Cartan involution; modulo a set of measure zero, $\overline{N}$ can be identified with the "maximal boundary" of the symmetric space G/K.

A function on G/K is called harmonic if it is annihilated by all G-invariant differential operators without constant term. Each bounded continuous function f on $\overline{N}$ canonically determines a harmonic function u on G/K by the formula

$$u(gK) = \int_{\overline{N}} P(g,x)f(x)dx,$$

where P, the Poisson kernel for G/K, is a function on $G \times \overline{N}$ which we shall identify more precisely below. Moreover, f is the "boundary value" of u in the following sense. Let H be any element of the positive Weyl chamber in $\mathfrak{a}$ (the Lie algebra of A), and let $h_t = \exp(tH) \in A$. Then

$$(8.17) \qquad \lim_{t\to\infty} u(xh_tK) = f(x) \qquad \text{for all } x \in \overline{N}.$$

Henceforth we fix an H in the positive Weyl chamber. We then have the following facts:

(i) Let $\eta = \max\{1/\alpha(H): \alpha$ is a positive restricted root$\}$. Then the formula

$$\delta_r x = h_t x h_t^{-1} \qquad (h_t = \exp(tH), \quad t = -\eta \log r)$$

defines a family of dilations on $\bar{N}$. (The factor of η is there to make the smallest exponent of the dilations equal to 1.)

(ii) With this notation, we have the following formula for the Poisson kernel:

$$P(xh_t, y) = \phi_{r^\eta}(y^{-1}x) \qquad (x,y \in \bar{N})$$

where $\phi(x) = \exp[-2\rho(H(x^{-1}))]$, 2ρ being the sum of the positive restricted roots. (Note: Korányi [2] confuses $\phi(x)$ with $\phi(x^{-1})$ in his Lemma 6.2.)

The limiting relation (8.17) therefore says precisely that $f*\phi_r(x) \to f(x)$ for all $x \in \bar{N}$ as $r \to 0$, when f is bounded and continuous. Our aim here is to study the maximal function M_ϕ on $\bar{N}$ with the object in mind of obtaining almost-everywhere convergence results for $f \in L^p$. We summarize the properties of ϕ which we shall need in the following proposition:

(8.18) PROPOSITION. (a) $\phi(x) > 0$ _for all_ x, _and_ $\int \phi(x)dx = 1$.

(b) $\phi(rx)$ _is a nonincreasing function of_ r _for each_ x.

(c) _There exist_ $C > 0$, $b \geq 0$ _such that_ $|Y_j\phi(x)| \leq C(1+|x|)^b$ _for_ $1 \leq j \leq n$.

(d) *There exists* $\varepsilon > 0$ *such that* $\int (1+|x|)^{\varepsilon}\phi(x)dx < \infty$.

(e) *There exists* $C > 0$ *such that* $\phi(xy) \leq C\phi(x)$ *for all* $x,y \in \overline{N}$ *with* $|y| \leq 1$.

Proof: The following facts about ϕ are due respectively to Knapp and Williamson [1] (Proposition 5.1) and Harish-Chandra [1] (Corollary to Lemma 45):

(i) $\phi = (1+\Sigma_1^k Q_j)^{-1}$ where each Q_j is a nonnegative homogeneous polynomial of positive degree, and the only $x \in \overline{N}$ such that $Q_j(x) = 0$ for all j is $x = 0$.

(ii) There exists $\varepsilon > 0$ such that $\int \phi(x)^{1-\varepsilon}dx < \infty$.

Assertions (a), (b), and (c) follow immediately from (i) and the fact that $f*\phi_r \to f$ as $r \to 0$ for $f \in C_0$. Also, (i) implies that $|\phi(x)| \leq C(1+|x|)^{-1}$, so by (ii),

$$\int (1+|x|)^{\varepsilon}\phi(x)dx \leq \sup_x[\phi(x)^{\varepsilon}(1+|x|)^{\varepsilon}] \int \phi(x)^{1-\varepsilon}dx < \infty.$$

which proves (d). Finally, to prove (e), suppose $g_1, g_2 \in G$. Then $H(g_1g_2) = H(g_1\kappa(g_2)) + H(g_2)$ (because A normalizes N). Since $\kappa(g_2)$ lies in the compact set K, it follows that for any compact $V \subset G$ there exists $C > 0$ such that

$$|H(g_1g_2) - H(g_2)| \leq C \quad \text{for} \quad g_1 \in V,\ g_2 \in G.$$

But $\phi(x) = \exp[-2\rho(H(x^{-1}))]$, so we obtain the desired estimate by taking $g_1 = y^{-1}$, $g_2 = x^{-1}$. #

Our main result is the following:

(8.19) THEOREM. *Let* ϕ *be a function on* $\overline{N}$ *satisfying the conditions in Proposition* 8.18 (*which we shall refer to as properties* (a) *through* (e)). *Then there exists* $p_0 < 1$ *such that the maximal operator* M_ϕ *is bounded from* H^p *to* L^p *for* $p > p_0$.

The proof is rather lengthy, and we begin by making some simple reductions. First, since $\phi \in L^1$, M_ϕ is trivially bounded on L^∞. By Theorem 3.34, then, it will suffice to consider $p \leq 1$, and for this it is enough to show that there exist $p_0 < 1$ and $C_p > 0$ for $p > p_0$ such that $\|M_\phi f\|_p \leq C_p$ whenever $p > p_0$ and f is a $(p,\infty,0)$-atom. (We shall need no higher moment conditions on f; however, see the remarks following Corollary 8.25.) Moreover, by composing with translations and dilations we may assume that f is associated to $B(1,0)$. In short, f will henceforth denote a function supported in $B(1,0)$ such that $\|f\|_\infty \leq 1$ and $\int f(x)dx = 0$. Also, we fix $J \in \mathbb{N}$ large enough so that $2^J \geq 3\gamma\beta$. Since $\|M_\phi f\|_\infty \leq \|f\|_\infty \|\phi\|_1 \leq 1$, we have

$$(8.20) \qquad \int_{|x|<2^J} (M_\phi f(x))^p dx \leq 2^{JQ} = \text{constant},$$

so we need only estimate the integral of $(M_\phi f(x))^p$ over the set where $|x| \geq 2^J$.

Let $x \to \langle x \rangle$ be a C^∞ function on G such that $\langle x \rangle = |x|$ when $|x| \geq 2$ and $1 \leq \langle x \rangle \leq 1 + |x|$ when $|x| \leq 2$. For $z \in \mathbb{C}$ we set

$$\phi^z(x) = \langle x \rangle^z \phi(x), \qquad \phi^z_t = (\phi^z)_t, \quad M^z f = M_{\phi^z} f.$$

Thus we wish to estimate $M^0 f = M_\phi f$.

(8.21) LEMMA. *Let* $S = \{x : |x| = 1\}$ *and* σ = *the surface measure on* S *given by Proposition* 1.15, *and let* ε *be as in property* (d). *For* $x' \in S$, *set*

$$\Omega(x') = \int_0^\infty \phi(rx')r^{Q-1+\varepsilon}dr.$$

Then $\Omega \in L^1(S,\sigma)$, *and there exists* $C > 0$ *such that*

$$\phi(x) \leq C(1+|x|)^{-Q-\varepsilon}\Omega(x') \qquad \text{for} \quad x \neq 0,\ x' = x/|x|.$$

Proof: By property (d) we have

$$\int_S \Omega(x')d\sigma(x') = \int_S \int_0^\infty \phi(rx')r^{Q-1+\varepsilon}drd\sigma(x') = \int_{\overline{N}} \phi(x)|x|^\varepsilon dx < \infty.$$

Also, by property (b), if $x = rx'$ $(r > 0,\ x' \in S)$,

$$(8.22) \qquad \Omega(x') \geq \int_0^r \phi(sx')s^{Q-1+\varepsilon}ds \geq \phi(rx')\int_0^r s^{Q-1+\varepsilon}ds = (Q+\varepsilon)^{-1}|x|^{Q+\varepsilon}\phi(x).$$

Thus for $|x| \geq 1$,

$$\phi(x) \leq (Q+\varepsilon)|x|^{-Q-\varepsilon}\Omega(x') \leq 2^{Q+\varepsilon}(Q+\varepsilon)(1+|x|)^{-Q-\varepsilon}\Omega(x').$$

On the other hand, if we set $\delta = \inf_{|x'|=1}\phi(x')$ we have $\delta > 0$ (by property (a)) and $\Omega(x') \geq \delta(Q+\varepsilon)$ (by (8.22)), so for $|x| \leq 1$,

$$\phi(x) \leq \phi(0) \leq \phi(0)2^{Q+\varepsilon}(Q+\varepsilon)\delta^{-1}(1+|x|)^{-Q-\varepsilon}\Omega(x'). \quad \#$$

(8.23) LEMMA. *There exists* $C_0 > 0$, *independent of* f, *such that if* $\operatorname{Re} z \leq \varepsilon$ (*where* ε *is as in property* (d)) *and* $j \geq J$,

$$\int_{2^j \leq |x| \leq 2^{j+1}} M^z f(x)dx \leq C_0.$$

Proof: Suppose $|x^{-1}y| < t$. We observe that

$$f*\phi_t^z(y) = \int f(xw^{-1})\phi_t^z(wx^{-1}y)dw$$

and that if $\operatorname{Re} z \leq \varepsilon$,

$$|\phi_t^z(wx^{-1}y)| \leq t^{-Q}\phi(t^{-1}(wx^{-1}y)) < t^{-1}(wx^{-1}y) >^{\varepsilon} .$$

But by property (e) and Lemma 8.21,

$$\phi(t^{-1}(wx^{-1}y)) \leq C\phi(w/t) \leq C'(1+|w/t|)^{-Q-\varepsilon}\Omega(w') \qquad (w' = w/|w|),$$

while by Lemma 1.10,

$$\begin{aligned} < t^{-1}(wx^{-1}y) >^{\varepsilon} &\leq (1+|t^{-1}(wx^{-1}y)|)^{\varepsilon} \\ &\leq \gamma^{\varepsilon}(1+|w/t|)^{\varepsilon}(1+|(x^{-1}y)/t|)^{\varepsilon} \\ &\leq (2\gamma)^{\varepsilon}(1+|w/t|)^{\varepsilon}, \end{aligned}$$

so that

$$\begin{aligned} |\phi_t^z(wx^{-1}y)| &\leq Ct^{-Q}(1+|w/t|)^{-Q}\Omega(w') \\ &\leq C|w|^{-Q}\Omega(w'). \end{aligned}$$

Therefore,

$$\begin{aligned} M^z f(x) &\leq C\int |f(xw^{-1})|\,|w|^{-Q}\Omega(w')dw \\ &\leq C\int_{|xw^{-1}|\leq 1} |w|^{-Q}\Omega(w')dw. \end{aligned}$$

However, if $j \geq J$, $2^j \leq |x| \leq 2^{j+1}$, and $|xw^{-1}| \leq 1$ we have

$$2^{j-1} \leq 2^j - \gamma \leq |w| \leq 2^{j+1} + \gamma \leq 2^{j+2}.$$

Hence, if χ denotes the characteristic function of $B(1,0)$,

$$M^z f(x) \leq C \int_{2^{j-1} \leq |w| \leq 2^{j+2}} \chi(xw^{-1}) |w|^{-Q} \Omega(w') dw,$$

and thus

$$\int_{2^j \leq |x| \leq 2^{j+1}} M^z f(x) dx \leq C \int_{\overline{N}} \int_{2^{j-1} \leq |w| \leq 2^{j+2}} \chi(xw^{-1}) |w|^{-Q} \Omega(w') dw dx$$

$$= C \int_{2^{j-1} \leq |w| \leq 2^{j+2}} |w|^{-Q} \Omega(w') dw$$

$$= C \int_{2^{j-1}}^{2^{j+2}} r^{-1} dr \int_S \Omega(w') d\sigma(w')$$

$$= C \log 8 \int_S \Omega(w') d\sigma(w'),$$

and this is finite by Lemma 8.21. #

(8.24) LEMMA. <u>Let</u> b <u>be as in property</u> (c). <u>There exists</u> $C_1 > 0$, <u>independent of</u> f, <u>such that if</u> $\operatorname{Re} z = -(Q+b+\bar{d})$ <u>and</u> $j \geq J$,

$$\int_{2^j \leq |x| \leq 2^{j+1}} M^z f(x) dx \leq C_1 |z| 2^{-j}.$$

Proof: Since $\langle x \rangle$ is homogeneous of degree 1 for large x, for $1 \leq i \leq n$, $v \in \bar{N}$, and $\operatorname{Re} z = -(Q+b+\bar{d})$ we have

$$|Y_i \phi^z(v)| \leq |(Y_i \langle v \rangle^z)\phi(v)| + |\langle v \rangle^z Y_i \phi(v)|$$

$$\leq C|z|(1+|v|)^{-Q-b-\bar{d}-1} + C(1+|v|)^{-Q-\bar{d}}$$

$$\leq C'|z|(1+|v|)^{-Q-d_i}.$$

Hence

$$|Y_i(\phi_t^z)(v)| = t^{-Q-d_i}|(Y_i\phi^z)(v/t)| \leq C'|z|(t+|v|)^{-Q-d_i},$$

so by the mean value theorem (1.33),

$$|\phi_t^z(w^{-1}y) - \phi_t^z(y)| \leq C|z|\Sigma_1^n |w|^{d_i} \sup_{|v| \leq \beta|w|} (t+|vy|)^{-Q-d_i}.$$

Now suppose $|x| \geq 2^J$ ($\geq 3\gamma\beta$), $|x^{-1}y| < t$, and $|w| \leq 1$. Then $|v| \leq \beta|w|$ implies $|vx| \geq 2|x|/3$, so if $t \leq |x|/3\gamma$,

$$t + |vy| \geq |vy| \geq |vx| - \gamma|x^{-1}y| \geq (2|x|/3) - (|x|/3) = |x|/3,$$

whereas if $t > |x|/3\gamma$,

$$t + |vy| \geq t \geq |x|/3\gamma.$$

Thus, for all $t > 0$,

$$|\phi_t^z(w^{-1}y) - \phi_t^z(y)| \leq C|z|\Sigma_1^n |w|^{d_i}|x|^{-Q-d_i}.$$

Finally, since $\int f = 0$, for $|x| \geq 2^J$ we have

$$M^z f(x) = \sup_{|x^{-1}y|<t<\infty} \left|\int f(w)[\phi_t^z(w^{-1}y) - \phi_t^z(w)]dw\right|$$

$$\leq C|z|\Sigma_1^n |x|^{-Q-d_i} \int_{|w|\leq 1} |w|^{d_i} dw$$

$$\leq C'|z||x|^{-Q-1},$$

and hence for $j \geq J$,

$$\int_{2^j\leq|x|\leq 2^{j+1}} M^z f(x)dx \leq C_1|z| \int_{2^j}^{2^{j+1}} r^{-2}dr \leq C_1|z|2^{-j}. \quad \#$$

Proof of Theorem 8.19: For $j \geq J$, let

$$X_j = L^1(\{x : 2^j \leq |x| \leq 2^{j+1}\}, dx),$$

and for any measurable $\tau : G \to (0,\infty)$ and any measurable $\eta : G \to G$ such that $|x^{-1}\eta(x)| < \tau(x)$ for all $x \in G$, set

$$(F_{\tau,\eta}(z))(x) = (z-1-\varepsilon)^{-1} f*\phi^z_{\tau(x)}(\eta(x)).$$

Lemma 8.23 shows that $F_{\tau,\eta}(z)$ is an analytic X_j-valued function of z in the half-plane $\mathrm{Re}\, z < \varepsilon$ which is continuous up to the boundary and bounded in norm by C_0. Also, by Lemma 8.24,

$$\|F_{\tau,\eta}(z)\|_{X_j} \leq C_1 2^{-j} \qquad \text{for } \mathrm{Re}\, z = -(Q+b+\bar{d}).$$

Therefore, by the three lines lemma (cf. Stein and Weiss [2], p. 180),

$$\|F_{\tau,\eta}(0)\|_{X_j} \leq C_0^{1-\theta} C_1^{\theta} 2^{-j\theta}, \quad \text{where} \quad \theta = \varepsilon/(Q+b+\bar{d}+\varepsilon).$$

Since this is true for all τ,η, we have

$$\int_{2^j \leq |x| \leq 2^{j+1}} M^0 f(x)dx \leq C2^{-j\theta}.$$

Let us now take $p_0 = Q/(Q+\theta)$: then, for $p_0 < p \leq 1$,

$$\int_{|x|\geq 2^J} (M^0 f(x))^p dx = \Sigma_J^\infty \int_{2^j \leq |x| \leq 2^{j+1}} (M^0 f(x))^p dx$$

$$\leq \Sigma_J^\infty \Big[\int_{2^j \leq |x| \leq 2^{j+1}} M^0 f(x)dx\Big]^p \Big[\int_{2^j \leq |x| \leq 2^{j+1}} dx\Big]^{1-p}$$

$$\leq C\Sigma_J^\infty 2^{-j\theta p + jQ(1-p)} = C' < \infty.$$

Combining this with (8.20), we are done. #

(8.25) COROLLARY. (a) <u>The maximal operator</u> M_ϕ <u>is weak type</u> (1,1).

(b) <u>If</u> $f \in L^p$, $1 \leq p \leq \infty$, <u>then</u>

$$\lim_{|x^{-1}y| < t \to 0} f * \phi_t(y) = f(x) \qquad \text{for almost every } x \in G.$$

<u>Proof</u>: (a) follows from Theorem 3.37, and (b) follows from (a) by the argument used to prove Theorem 2.6. #

Our proof of Theorem 8.19 does not yield the optimal value of p_0. The estimate in Lemma 8.24 can be improved by assuming that f has many vanishing moments, using estimates on higher order derivatives of ϕ (which are valid for the Poisson kernel: cf. the proof of Proposition 8.18). and replacing the mean value theorem by a higher order Taylor inequality. The improvement in the final result is not great, however, for one must also decrease $\operatorname{Re} z$ in the hypothesis of Lemma 8.24, with a corresponding decrease in θ in the final argument. We leave it to the interested reader to work out the details. What is most important is that, in general, one cannot take $p_0 = 0$. That this is so can be seen by considering the Poisson kernel for the product of two half-planes. Here $G = (SL(2,\mathbb{R}))^2$, $\bar{N} = \mathbb{R}^2$, the dilations on $\bar{N}$ corresponding to the various elements of the positive Weyl chamber are given by

$$\delta_r(x_1,x_2) = (r^{d_1}x_1, r^{d_2}x_2) \qquad (\min(d_1,d_2) = 1),$$

and the Poisson kernel is given by

$$\phi(x_1,x_2) = [\pi^2(1+x_1^2)(1+x_2^2)]^{-1}.$$

Let f_1 be a bounded function $\mathbb{R}$ with compact support and as many vanishing moments as one pleases, and let f_2 be the characteristic function of $[0,1]$. Then $f(x_1,x_2) = f_1(x_1)f_2(x_2)$ has as many vanishing moments as f_1, hence is a constant multiple of an atom on $\mathbb{R}^2$. Moreover,

$$f*\phi(x_1,x_2) = g(x_1)[\arctan(x_2)-\arctan(x_2-1)], \quad \text{where}$$

$$g(x_1) = \frac{1}{\pi^2}\int_{-\infty}^{\infty} \frac{f_1(x_1-y)}{1+y^2}\,dy.$$

But $\arctan(x_2)-\arctan(x_2-1)$ is asymptotic to x_2^{-2} as $x_2 \to \pm\infty$, so

$$M_\phi f(x_1,x_2) \geq |f*\phi(x_1,x_2)| \geq C|g(x_1)|(1+x_2^2)^{-1},$$

which fails to lie in L^p when $p \leq 1/2$.

Notes and References

The analogue of Theorem 8.8 for harmonic functions on $\mathbb{R}^n \times (0,\infty)$ and the main results of Section B for the classical Poisson kernel on $\mathbb{R}^n$ are due to Fefferman and Stein [1]. Some of the corresponding results for Poisson-Szegö integrals on the Heisenberg group have been obtained by Geller [2]. Theorem 8.19 is new, but Corollary 8.25 was originally proved by Stein [5] without recourse to H^p theory.

BIBLIOGRAPHY

J. Bergh and J. Löfström

1. Interpolation Spaces (Grund. Math. Wiss., no. 223), Springer-Verlag, Berlin, 1976.

J. M. Bony

1. Principe du maximum, inégalité de Harnack, et unicité du problème de Cauchy pour les opérateurs elliptiques dégénérés, Ann. Inst. Fourier 19 (1) (1969), 277-304.

N. Bourbaki

1. Groupes et Algèbres de Lie, Chap. I-III (Eléments de Math., Fasc. 26 et 37), Hermann, Paris, 1960 and 1972.

D. L. Burkholder, R. F. Gundy, and M. L. Silverstein

1. A maximal function characterization of the class H^p, Trans. Amer. Math. Soc. 157 (1971), 137-153.

A. P. Calderón

1. An atomic decomposition of distributions in parabolic H^p spaces, Advances in Math. 25 (1977), 216-225.
2. Commutators of singular integral operators, Proc. Nat. Acad. Sci. USA 53 (1965), 1092-1099.

A. P. Calderón and A. Torchinsky

1. Parabolic maximal functions associated with a distribution, Advances in Math. 16 (1975), 1-64.
2. Parabolic maximal functions associated with a distribution, II, Advances in Math. 24 (1977), 101-171.

S. Campanato

1. Proprietà di una famiglia di spazi funzionali, Ann. Scuola Norm. Sup. Pisa 18 (1964), 137-160.

L. Carleson

1. Two remarks on H^1 and BMO, Advances in Math. 22 (1976), 269-277.

S. Y. A. Chang and R. Fefferman

1. A continuous version of duality of H^1 and BMO on the bidisc, Ann. of Math. 112 (1980), 179-201.

J. A. Chao, J. E. Gilbert, and P. A. Tomas

1. Molecular decompositions in H^p theory, Rend. Circ. Mat. Palermo supplemento, 1 (1981), 115-119.

R. R. Coifman

1. A real variable characterization of H^p, Studia Math. 51 (1974), 269-274.

R. R. Coifman and Y. Meyer
1. Au delà des operateurs pseudo-differentiels, Astérisque 57 (1978), 1-185.

R. R. Coifman and R. Rochberg
1. Representation theorems for holomorphic and harmonic functions in L^p, Astérisque 77 (1980), 11-66.

R. R. Coifman and G. Weiss
1. Analyse Harmonique Non-commutative sur Certains Espaces Homogènes (Lecture Notes in Math., no. 242), Springer-Verlag, Berlin, 1971.
2. Extensions of Hardy spaces and their use in analysis, Bull. Amer. Math. Soc. 83 (1977), 569-645.

L. de Michele and G. Mauceri
1. L^p multipliers on the Heisenberg group, Michigan Math. J. 26 (1979), 361-371.

J. Dixmier and P. Malliavin
1. Factorisations de fonctions et de vecteurs indéfiniment différentiables, Bull. Sci. Math. 102 (1978), 305-330.

P. L. Duren
1. Theory of H^p spaces, Academic Press, N.Y., 1971.

P. L. Duren, B. W. Romberg, and A. L. Shields
1. Linear functionals on H^p spaces with $0 < p < 1$, J. Reine Angew. Math. 238 (1969), 32-60.

J. L. Dyer
1. A nilpotent Lie algebra with nilpotent automorphism group, Bull. Amer Math. Soc. 76 (1970), 52-56.

E. B. Fabes and C. E. Kenig
1. On the Hardy space H^1 of a C^1 domain, to appear in Arkiv för Mat.

E. B. Fabes and N. M. Rivière
1. Singular integrals with mixed homogeneity, Studia Math. 27 (1966), 19-38

C. Fefferman
1. Characterizations of bounded mean oscillation, Bull. Amer. Math. Soc. 77 (1971), 587-588.
2. Harmonic analysis and H^p spaces, pp. 38-75 in Studies in Harmonic Analysis (ed. by J. M. Ash), Mathematical Association of America, 1976.

C. Fefferman, N. M. Rivière, and Y. Sagher
1. Interpolation between H^p spaces: the real method, Trans. Amer. Math. Soc. 191 (1974), 75-81

C. Fefferman and E. M. Stein
1. H^p spaces of several variables, Acta Math. 129 (1972), 137-193.

G. B. Folland

1. Subelliptic estimates and function spaces on nilpotent Lie groups, *Arkiv för Math.* 13 (1975), 161-207.
2. Applications of analysis on nilpotent groups to partial differential equations, *Bull Amer. Math. Soc.* 83 (1977), 912-930.
3. Lipschitz classes and Poisson integrals on stratified groups, *Studia Math.* 66 (1979), 37-55.

G. B. Folland and E. M. Stein

1. Estimates for the ∂_b complex and analysis on the Heisenberg group, *Comm. Pure Appl. Math.* 27 (1974), 429-522.

J. García-Cuerva

1. Weighted H^p spaces, *Dissertationes Math.*, no. 162 (1979).

J. B. Garnett and R. H. Latter

1. The atomic decomposition for Hardy spaces in several complex variables, *Duke Math. J.* 45 (1978), 815-845.

D. Geller

1. Fourier analysis on the Heisenberg group, I: the Schwartz space, *J. Functional Analysis* 36 (1980), 205-254.
2. Some results in H^p theory for the Heisenberg group, *Duke Math. J.* 47 (1980), 365-390.

D. Goldberg

1. A local version of real Hardy spaces, *Duke Math. J.* 46 (1979), 27-42.

R. W. Goodman

1. *Nilpotent Lie Groups: Structure and Applications to Analysis* (Lecture Notes in Math., no. 562), Springer-Verlag, Berlin, 1976.
2. Singular integral operators on nilpotent Lie groups, *Arkiv för Math.* 18 (1980), 1-11.

P. C. Greiner and E. M. Stein

1. *Estimates for the ∂-Neumann Problem* (Math. Notes, no. 19), Princeton University Press, Princeton, N.J., 1977.

R. F. Gundy and E. M. Stein

1. H^p theory for the poly-disc, *Proc. Nat. Acad. Sci. USA* 76 (1979), 1026-1029.

Harish-Chandra

1. Spherical functions on a semisimple Lie group, I, *Amer. J. Math.* 80 (1958), 241-310.

B. Helffer and J. Nourrigat

1. Caractérisation des opérateurs hypoelliptiques homogènes invariants à gauche sur un groupe de Lie nilpotent gradué, *Comm. Partial Diff. Eq.* 4 (1979), 899-958.

S. Helgason

1. A duality for symmetric spaces with applications to group representations, _Advances in Math_. 5 (1970), 1-154.

M. L. Hemler

1. _The Molecular Theory of_ $H^{p,q,s}(\mathbb{H}^n)$, Ph.D. Dissertation, Washington University, St. Louis, 1980.

C. Herz

1. H_p spaces of martingales, $0 < p \leq 1$, _Zeit. Warschein_. 28 (1974), 189-205.

L. Hörmander

1. Hypoelliptic second-order differential equations, _Acta Math_. 119 (1967), 147-171.

A. Hulanicki

1. Subalgebra of $L_1(G)$ associated with Laplacian on a Lie group, _Colloq. Math_. 31 (1974), 259-287.
2. Commutative subalgebra of $L^1(G)$ associated with a subelliptic operator on a Lie group G, _Bull. Amer. Math. Soc_. 81 (1975), 121-124.

A. Hulanicki and E. M. Stein

1. Marcinkiewicz multiplier theorem for stratified groups (manuscript).

G. A. Hunt

1. Semigroups of measures on Lie groups, _Trans. Amer. Math. Soc_. 81 (1956), 264-293.

S. Igari

1. An extension of the interpolation theorem of Marcinkiewicz II, _Tohoku Math. J_. 15 (1963), 343-358.

S. Janson

1. Generalizations of Lipschitz spaces and an application to Hardy spaces and bounded mean oscillation, _Duke Math. Jour_. 47 (1980), 959-982.

J. W. Jenkins

1. Dilations and gauges on nilpotent Lie groups, _Colloq. Math_. 41 (1979), 95-101.

D. S. Jerison

1. The Dirichlet problem for the Kohn Laplacian on the Heisenberg group (preprint).

F. John and L. Nirenberg

1. On functions of bounded mean oscillation, _Comm. Pure Appl. Math_. 14 (1961), 415-426.

P. W. Jones
1. Carleson measures and the Fefferman-Stein decomposition of BMO($\mathbb{R}$), *Ann. of Math.* 111 (1980), 197-208.

A. Kaplan and R. Putz
1. Boundary behaviour of harmonic forms on a rank one symmetric space, *Trans. Amer. Math. Soc.* 231 (1977), 369-384.

A. W. Knapp and E. M. Stein
1. Intertwining operators for semisimple groups, *Ann. of Math.* 93 (1971), 489-578.

A. W. Knapp and R. E. Williamson
1. Poisson integrals and semisimple groups, *J. Analyse Math.* 24 (1971), 53-76.

P. Koosis
1. *Introduction to H_p Spaces* (London Math. Soc. Lecture Notes, no. 40), Cambridge University Press, Cambridge, U.K. 1980.

A. Korányi
1. Harmonic functions on hermitian hyperbolic space, *Trans. Amer. Math. Soc.* 135 (1969), 507-516.
2. Harmonic functions on symmetric spaces, pp. 379-412 in *Symmetric Spaces* (ed. by W. Boothby and G. Weiss), Marcel Dekker, New York, 1972.

A. Korányi and S. Vági
1. Singular integrals in homogeneous spaces and some problems of classical analysis, *Ann. Scuola Norm. Sup. Pisa* 25 (1971), 575-648.

S. G. Krantz
1. Structure and interpolation theorems for certain Lipschitz spaces and estimates for the $\bar{\partial}$ equation, *Duke Math. J.* 43 (1976), 417-439.
2. Geometric Lipschitz spaces and applications to complex function theory and nilpotent groups, *J. Functional Analysis* 34 (1979), 456-471.
3. Analysis on the Heisenberg group and estimates for functions in Hardy classes of several complex variables, *Math. Ann.* 244 (1979), 243-26.

R. H. Latter
1. A characterization of $H^p(\mathbb{R}^n)$ in terms of atoms, *Studia Math.* 62 (1978), 93-101.

R. H. Latter and A. Uchiyama
1. The atomic decomposition theorem for parabolic H^p spaces, *Trans. Amer. Math. Soc.* 253 (1979), 391-398.

R. A. Macías
1. Interpolation Theorems on Generalized Hardy Spaces, Ph.D. Dissertation, Washington University, St. Louis, 1974.

R. A. Macías and C. Segovia
1. Lipschitz functions on spaces of homogeneous type, Advances in Math. 33 (1979), 257-270.
2. A decomposition into atoms of distributions on spaces of homogeneous type, Advances in Math. 33 (1979), 271-309.

B. Marshall, W. Strauss, and S. Wainger
1. L^p-L^q estimates for the Klein-Gordon equation, Jour. de Math. Pures et Appl. 59 (1980), 417-440.

G. Mauceri, M. Picardello, and F. Ricci
1. A Hardy space associated with twisted convolution, Advances in Math. 39 (1981), 270-288.

N. G. Meyers
1. Mean oscillation over cubes and Hölder continuity, Proc. Amer. Math. Soc. 15 (1964), 717-721.

K. G. Miller
1. Parametrices for hypoelliptic operators on step two nilpotent Lie groups, Comm. Partial Diff. Eq. 5 (1980), 1153-1184.

A. Miyachi
1. On some Fourier multipliers for $H^p(\mathbb{R}^n)$, Jour. Fac. Sci. Univ. Tokyo 27 (1980), 157-179.

A. Nagel and E. M. Stein
1. Lectures on Pseudo-differential Operators (Math. Notes, no. 24), Princeton University Press, Princeton, N.J., 1979.

E. Nelson and W. F. Stinespring
1. Representation of elliptic operators in an enveloping algebra, Amer. J. Math. 81 (1959), 547-560.

J. Peral
1. L^p estimates for the wave equations, Jour. of Funct. Analysis 36 (1980), 114-145.

F. Ricci
1. Calderón-Zygmund kernels on nilpotent Lie groups (preprint).

C. Rockland
1. Hypoellipticity for the Heisenberg group, Trans. Amer. Math. Soc. 240 (1978), 1-52.

L. P. Rothschild and E. M. Stein
1. Hypoelliptic differential operators and nilpotent groups, Acta Math. 137 (1976), 247-320.

K. Saka
1. Besov spaces and Sobolev spaces on a nilpotent Lie group, Tohoku Math. J. 31 (1979), 383-437.

C. Segovia
1. On the area function of Lusin, Studia Math. 33 (1969), 312-343.

P. Sjolin
1. An H^p inequality for strongly singular integrals, Math. Zeit. 165 (1979), 231-238.

E. M. Stein
1. On the functions of Littlewood-Paley, Lusin, and Marcinkiewicz, Trans. Amer. Math. Soc. 88 (1958), 430-466.
2. Singular Integrals and Differentiability Properties of Functions, Princeton University Press, Princeton, N.J., 1970.
3. Topics in Harmonic Analysis Related to the Littlewood-Paley Theory (Ann. of Math. Studies, no. 63), Princeton University Press, Princeton, N.J., 1970.
4. Some problems in harmonic analysis suggested by symmetric spaces and semisimple groups, Proc. Internat. Congress Math. Nice (1970), vol. I., 173-189.
5. Maximal functions: Poisson integrals on symmetric spaces, Proc. Nat. Acad. Sci. USA 73 (1976), 2547-2549.
6. Lectures on H^p Spaces, unpublished notes by D. Jerison, Princeton University, 1977.

E. M. Stein, M. H. Taibleson, and G. Weiss
1. Weak type estimates for maximal operators on certain H^p spaces, Rend. Circ. Mat. Palermo, supplemento, 1 (1981), 81-97.

E. M. Stein and G. Weiss
1. On the theory of harmonic functions of several variables, I. The theory of H^p spaces, Acta Math. 103 (1960), 25-62.
2. Introduction to Fourier Analysis on Euclidean Spaces, Princeton University Press, Princeton, N.J., 1971.

E. M. Stein and A. Zygmund
1. Boundedness of translation invariant operators on Hölder spaces and L^p spaces, Ann. of Math. 85 (1967), 337-349.

R. S. Strichartz
1. Singular integrals on nilpotent Lie groups, Proc. Amer. Math. Soc. 53 (1975), 367-374.

J. O. Strömberg
1. Bounded mean oscillation with Orlicz norms and duality of Hardy spaces, Indiana U. Math. J. 28 (1979), 511-544.

J. O. Strömberg and A. Torchinsky
1. Weights, sharp maximal functions, and Hardy spaces, Bull. Amer. Math. Soc. 3 (1980), 1053-1056.

M. H. Taibleson and G. Weiss
1. The molecular characterization of certain Hardy spaces, Astérisque 77 (1980), 67-149.

F. Trèves
1. Topological Vector Spaces, Distributions, and Kernels, Academic Press, New York, 1967.

A. Uchiyama
1. A maximal function characterization of H^p on the space of homogeneous type, Trans. Amer. Math. Soc. 262 (1980), 579-592.
2. Grand maximal functions and radial maximal functions (preprint).
3. A constructive proof of the Fefferman-Stein decomposition of $BMO(\mathbb{R}^n)$ (preprint).

T. Walsh
1. The dual of $H^p(\mathbb{R}^{n+1}_+)$ for $p < 1$, Canad. J. Math. 25 (1973), 567-577.

A. Zygmund
1. Trigonometric Series, Cambridge University Press, 1959.

Index of Terminology

Index of Notation

www.ingramcontent.com/pod-product-compliance
Ingram Content Group UK Ltd.
Pitfield, Milton Keynes, MK11 3LW, UK
UKHW020237161224
452563UK00006B/209

9 780691 083100